OOK

3

NELSON MATHS

WORKBOOK

YEARS 7–10

Megan Boltze
Deborah Smith
Robert Yen
Ilhea Yen

WORKSHEETS

PUZZLE SHEETS

HOMEWORK ASSIGNMENTS

NELSON
A Cengage Company

Nelson Maths Workbook 3
1st Edition
Megan Boltze
Deborah Smith
Robert Yen
Ilhea Yen
ISBN 9780170454537

Publisher: Robert Yen
Project editor: Alan Stewart
Editor: Anna Pang
Cover design: James Steer
Text design: Original text design by Alba Design, Adapted by James Steer
Project designer: James Steer
Permissions researcher: Corrina Gilbert
Production controller: Karen Young
Text illustrations: Cat MacInnes
Typeset by: MPS Limited

Any URLs contained in this publication were checked for currency during the production process. Note, however, that the publisher cannot vouch for the ongoing currency of URLs.

For product information and technology assistance,
in Australia call **1300 790 853**;
in New Zealand call **0800 449 725**

For permission to use material from this text or product, please email
aust.permissions@cengage.com

ISBN 978 0 17 045453 7

Cengage Learning Australia
Level 7, 80 Dorcas Street
South Melbourne, Victoria Australia 3205

Cengage Learning New Zealand
Unit 4B Rosedale Office Park
331 Rosedale Road, Albany, North Shore 0632, NZ

For learning solutions, visit **cengage.com.au**

Printed in China by 1010 Printing International Limited.
1 2 3 4 5 6 7 24 23 22 21 20

CONTENTS

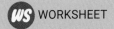 WORKSHEET   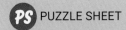 PUZZLE SHEET   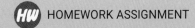 HOMEWORK ASSIGNMENT

CONTENTS

This 200-page workbook contains worksheets, puzzles, StartUp assignments and homework assignments written for the Australian Curriculum in Mathematics. It can be used as a valuable resource for teaching Year 9 mathematics, regardless of the textbook used in the classroom, and takes a wholistic approach to the curriculum, including some Year 8 and Year 10 work as well (there is also an Advanced edition of this workbook that contains some extension worksheets). This workbook is designed to be handy for homework, assessment, practice, revision, relief classes or 'catch-up' lessons.

Inside:

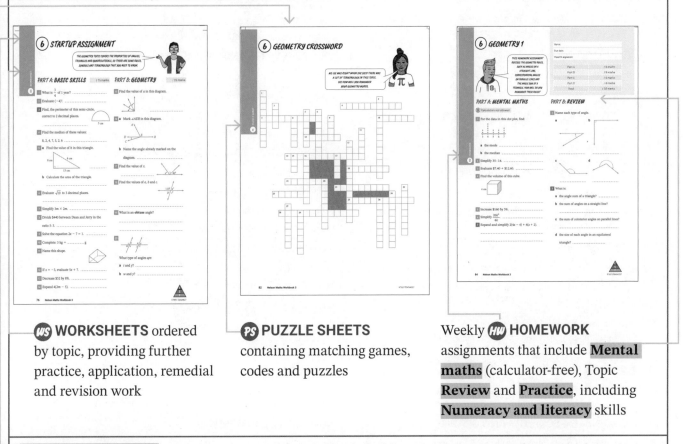

WS WORKSHEETS ordered by topic, providing further practice, application, remedial and revision work

PS PUZZLE SHEETS containing matching games, codes and puzzles

Weekly **HW HOMEWORK** assignments that include **Mental maths** (calculator-free), Topic **Review** and **Practice**, including **Numeracy and literacy** skills

StartUp assignments beginning each topic, revising skills from previous topics and prerequisite knowledge for the topic, including basic skills, review of a specific topic and a challenge problem

Word puzzles, such as a crossword or find-a-word, that reinforce the language of mathematics learned in the topic

The ideas and activities presented in this book were written by practising teachers and used successfully in the classroom.

Colour-coding of selected questions

Questions on most worksheets are graded by level of difficulty:

C — Complex
S — Standard
F — Foundation

CURRICULUM GRID

Topic skills	Australian curriculum strand and substrand
1 ALGEBRA	**NUMBER AND ALGEBRA**
From words to algebraic expressions, substitution, simplifying algebraic expressions, including algebraic fractions, expanding and factorising expressions, binomial products	Patterns and algebra
2 PYTHAGORAS' THEOREM	**MEASUREMENT AND GEOMETRY**
Finding an unknown side, testing for right-angled triangles, Pythagorean triads	Pythagoras and trigonometry
3 NUMERACY AND CALCULATION	**NUMBER AND ALGEBRA**
Integers, decimals, fractions, percentages, profit, discount, GST, simple interest, ratios and rates, converting rates, time differences	Real numbers Money and financial mathematics
4 TRIGONOMETRY	**MEASUREMENT AND GEOMETRY**
Trigonometric ratios, finding an unknown side, finding an unknown angle	Pythagoras and trigonometry
5 INDICES	**NUMBER AND ALGEBRA**
Index laws, the zero index, negative indices, significant figures, scientific notation	Real numbers Patterns and algebra
6 GEOMETRY	**MEASUREMENT AND GEOMETRY**
Angle geometry, triangle geometry, quadrilateral geometry	Geometric reasoning
7 EQUATIONS	**NUMBER AND ALGEBRA**
Equations with variables on both sides, with brackets, with algebraic fractions, equation problems, formulas	Patterns and algebra Linear and non-linear relationships
8 EARNING MONEY	**NUMBER AND ALGEBRA**
Wages, salaries, overtime, commission, piecework, leave loading, income tax, PAYG tax, net pay	Real numbers Money and financial mathematics
9 ANALYSING DATA	**STATISTICS AND PROBABILITY**
Mean, median, mode, range, histograms, stem-and-leaf plots, shape of a distribution, comparing data sets, sampling, types of data, bias and questionnaires	Data representation and interpretation
10 SURFACE AREA AND VOLUME	**MEASUREMENT AND GEOMETRY**
Metric units, perimeter, area, circles and sectors, surface areas and volumes of prisms and cylinders	Using units of measurement
11 COORDINATE GEOMETRY AND GRAPHS	**NUMBER AND ALGEBRA**
Length and midpoint of an interval, gradient of a line, graphing linear equations, $y = mx + c$, finding the equation of a line, solving linear equations graphically, direct proportion, graphing simple quadratic equations $y = ax^2 + c$ and circles	Real numbers Linear and non-linear relationships
12 PROBABILITY	**STATISTICS AND PROBABILITY**
Probability, relative frequency, Venn diagrams, two-way tables, two-step experiments	Chance
13 CONGRUENT AND SIMILAR FIGURES	**MEASUREMENT AND GEOMETRY**
Tests and proofs for congruent triangles, using congruence to prove geometrical properties, properties of similar figures, scale diagrams, tests for similar triangles	Geometric reasoning

MEET YOUR MATHS GUIDES ...

THIS WORKBOOK CONTAINS WORKSHEETS, PUZZLE SHEETS AND HOMEWORK ASSIGNMENTS

INTRODUCING MS LEE.

HI, I'VE BEEN TEACHING MATHS FOR OVER 20 YEARS

I BECAME GOOD AT MATHS THROUGH PRACTICE AND EFFORT

I WILL GUIDE YOU THROUGH THE WORKSHEETS

MATHS IS ABOUT MASTERING A COLLECTION OF SKILLS, AND I CAN HELP YOU DO THIS

THIS IS ZINA, A MATHS TUTOR AND MS LEE'S YEAR 12 STUDENT

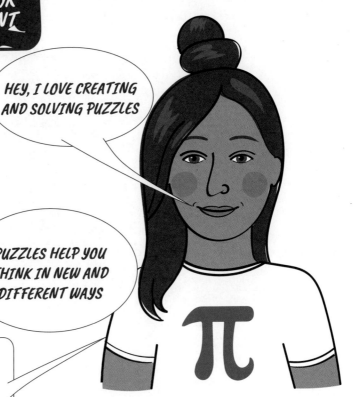

HEY, I LOVE CREATING AND SOLVING PUZZLES

NOT JUST MATHS PUZZLES BUT WORD PUZZLES TOO!

PUZZLES HELP YOU THINK IN NEW AND DIFFERENT WAYS

LET ME SHOW YOU HOW, AND YOU'LL GET SMARTER ALONG THE WAY

① STARTUP ASSIGNMENT 1

HERE ARE SOME SKILLS YOU NEED TO KNOW TO DO WELL IN MATHS.
PART A IS MIXED REVISION, PART B IS FOR LEARNING ALGEBRA.

PART A: BASIC SKILLS — / 15 marks

1 In which month does summer in Australia begin? _____

2 Evaluate $\dfrac{6.1 + 8.4}{5}$. _____

3 Draw a square pyramid.

4 Find the value of x in this diagram.

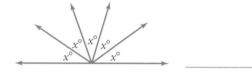

_____

5 Find $12\dfrac{1}{2}$ % of $90. _____

6 Name this shape.

_____

7 1.5 km = _____ m.

8 Half of $12 plus one-third of $12.

_____

9 What are parallel lines?

_____

_____

10 Simplify 48 : 36. _____

11 A square has a perimeter of 100 cm. What is its side length? _____

12 Write $\dfrac{4}{11}$ as a decimal. _____

13 If $x = 5$, evaluate $x^2 - x$. _____

14 What percentage of one hour is 48 minutes?

_____

15 Find the value of y in this triangle.

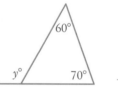

_____

PART B: ALGEBRA AND NUMBER

/ 25 marks

16 Simplify:

  a $5y - 9y$ _____

  b $m \times m \times m$ _____

  c $12x \div x$ _____

17 Evaluate:

  a $4 \times (-2) + 7$ _____

  b $-5 \times 5 + (-5) \times 1$ _____

18 Find:

  a the product of 8 and 4

_____

  b the highest common factor of 14 and 21

_____

  c the next odd number after 5 _____

19 Simplify $\dfrac{3}{15}$. _____

20 If $a = 3$ and $b = -1$, evaluate:

 a $2a + b$ _____

 b $b + 10$ _____

 c $3ab$ _____

21 Decrease 15 by 5. _____

22 Expand:

 a $3(p + 2)$ _____

 b $4(2k - 3)$ _____

23 Write an algebraic expression for twice n less 4.

_____

24 Complete this table using the rule $y = 2x - 5$.

x	7	3	0	2	−1
y					

25 Find the rule for this table.

_____

x	−1	0	1	2	3	4
y	−2	1	4	7	10	13

26 Complete this table by looking at the matchstick pattern below.

Triangle	1	2	3	4	21	100
Number of matches	3	5	7			

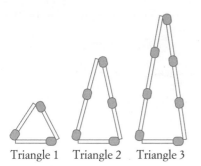

Triangle 1   Triangle 2   Triangle 3

PART C: CHALLENGE Bonus / 3 marks

How many squares can you see in this 3 × 3 grid? There are more than 9. Can you see 14?

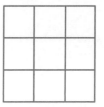

How many squares are there in a 4 × 4 grid?

_____

How many squares are there in a 5 × 5 grid?

_____

Calculate how many squares there are on a chess board (an 8 × 8 grid) by looking for a pattern in your previous answers.

_____

1 ALGEBRA REVIEW

THIS WORKSHEET REVISES THE ALGEBRA SKILLS FROM THIS TOPIC. ASK YOUR TEACHER OR ANOTHER STUDENT IF YOU GET STUCK.

1 Simplify each expression.

a $6u + 10 - u - 3$ _____

b $4xy + 3b - 2xy + b$ _____

c $n^2 + n^2$ _____

d $5t - 7t^2 + 4t + 2t^2$ _____

e $9m - m - 9 + 4$ _____

f $-4de + 7de + e$ _____

g $2x - y + 3y - x$ _____

h $2ab + 5b + 2ab - 5b$ _____

2 Write an algebraic expression for each statement.

a The cost in dollars of 4 tennis balls which cost $d each. _____

b The difference between 10 and k. _____

c The number of days in y weeks. _____

d Simone's age in t years if she is 14 this year.

_____

e The change in dollars from $100 after buying m tickets at $6 each. _____

f The average of 3 consecutive numbers beginning with n. _____

3 Simplify each expression.

a $7d \times 3e$ _____

b $2 \times p \times q \times p \times q$ _____

c $a^2 \times a$ _____

d $4r \times 4r$ _____

e $kp \times pr$ _____

f $3b^2 \times 3b^2$ _____

g $6m^4 \times 2m^2$ _____

h $18bc^2 \div 3bc$ _____

i $6x \div 3xy$ _____

j $\dfrac{15m}{10n}$ _____

k $\dfrac{20u}{2u}$ _____

l $\dfrac{4a^2}{8}$ _____

4 If $x = 4$ and $w = -2$, evaluate each expression.

a $x - w$ _____

b $3x + w$ _____

c $5w^2$ _____

d $\dfrac{w}{x}$ _____

e $4w - 2x + 5$ _____

f $x^2 \div w$ _____

5 The area of a trapezium is given by the formula $A = \dfrac{1}{2}(a + b)h$. Find the area of a trapezium that has parallel sides of length 3 cm and 7 cm and a perpendicular height of 5 cm.

_____

_____

_____

6 Expand and simplify:

a $x(x - 3)$ _____

b $-4(2x + 6)$ _____

c $u(w - 4)$ _____

d $-2(6 - 5b)$ _____

e $2(y - 5) + 4(y + 7)$ _____

f $7 - 2(r + 1)$ _____

g $a(a + 5) + 3(a - 2)$ _____

h $2(p + q) - (2p + q)$ _____

7 Write an algebraic expression for:

a the perimeter of this rectangle _____

$2x$

$x - 7$

b the area of the rectangle. _____

8 From a tower of height h metres, you can see for a distance of d km, where d is given by the formula $d = 8\sqrt{\dfrac{h}{5}}$. How far could you see from the top of Sydney Tower, which has a height of 305 metres? Answer to the nearest metre.

_____

9 Simplify each expression.

a $\dfrac{5x}{10} - \dfrac{3x}{10}$ _____

b $\dfrac{2u}{5} + \dfrac{u}{3}$ _____

c $\dfrac{3m}{8} + \dfrac{m}{2}$ _____

d $\dfrac{a}{2} - \dfrac{a}{3}$ _____

e $\dfrac{17}{4d} - \dfrac{3}{d}$ _____

f $\dfrac{2w}{3} \times \dfrac{3y}{10}$ _____

g $\dfrac{24p}{5q} \times \dfrac{15q}{3}$ _____

h $\dfrac{3r}{2x} \div \dfrac{10}{6x}$ _____

i $\dfrac{16ab}{15c} \div \dfrac{2b}{5}$ _____

1 ALGEBRA CROSSWORD

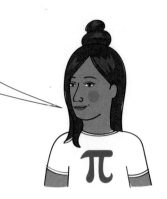

THIS CROSSWORD HAS MANY OF THE IMPORTANT KEY WORDS FROM THE ALGEBRA TOPIC. SOME OF THE CLUES ARE TRICKY.

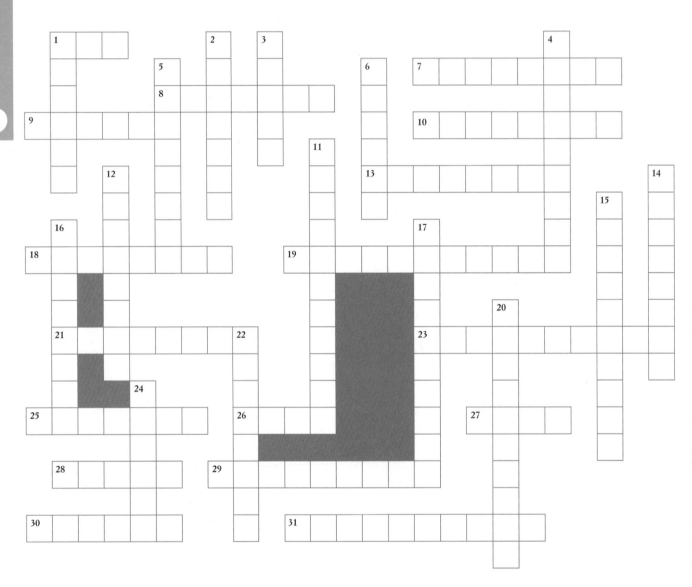

Clues across

1 The answer to an addition

7 Another name for 'parentheses'

8 Branch of mathematics that studies numerical patterns

9 A whole number that divides evenly into a given number

10 A number or expression written in the form $\frac{a}{b}$

13 To make smaller

18 The answer to a division

19 The bottom number in a fraction

21 To make bigger

23 A statement involving variables is an algebraic e_____

25 Brackets are an example of grouping _____

26 $3ab$, ab and $7ab$ are examples of _____ terms

27 Another 'R' word for relationship

28 $3k$, $4t$ and z are examples of algebraic _____

29 To take out a common factor

30 To remove the brackets from $3(x - 6)$

31 To divide by a fraction, multiply by its _____

Clues down

1 1, 4, 9 and 16 are examples of _____ numbers

2 When factorising, find the _____ common factor

3 A 2-row grid displaying a number pattern is called a _____ of values

4 The total length of a shape's boundary

5 A letter or symbol that stands for a number

6 ÷ means to _____

11 Numbers that follow one another, such as 7, 8, 9

12 The opposite of add

14 ( ) and [ ] are examples of g_____ symbols

15 To replace a variable with a number

16 You do this to find the product of 2 numbers

17 The answer to a subtraction

20 Another name for a variable

22 To find the value of an expression

24 HCF stands for highest _____ factor

1 ALGEBRA 1

HEY, I'M MITCH, ZINA'S FRIEND. I LIKE SPORT SO SKILL PRACTICE IS IMPORTANT TO ME. I'LL BE YOUR HOMEWORK COACH, TAKING YOU ON A PROGRAM OF DIFFERENT DRILLS.

HOMEWORK

PART A: MENTAL MATHS

🚫 Calculators not allowed

1 Write the formula for the circumference of a circle with radius r.

2 Evaluate:

a $122 \div 4$

b $15 \times 2 \times 8$

c $25 \div \dfrac{5}{6}$

3 Find the median of 6, 3, 2, 6, 5, 4, 6.

4 Find the volume of this prism.

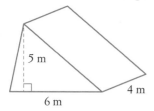

5 m
6 m
4 m

5 Simplify $\dfrac{9ab}{3b}$.

6 How many degrees in a straight angle?

PART B: REVIEW

1 Simplify:

a $6 \times y \times 3$ _____

b $8d - 5d - 4e + 9e$ _____

c $-5h \times 7m$ _____

2 Write $5m^2np$ in expanded form.

3 Write an algebraic expression for the number 6 more than Q.

4 Circle the like terms:

$4a^2b$, $3ab$, $6a^2$, $3a^2b$, ba^2.

5 Expand $9y(y - 2)$.

6 If $p = -7$ and $r = 4$, evaluate $2p + 3r - 9$.

PART C: PRACTICE

 › From words to algebraic expressions
› Substitution
› Adding and subtracting terms
› Multiplying and dividing terms

1 If $x = 10$ and $y = -3$, evaluate:

a $4y^2 + 5x$ _____

b $\dfrac{2(x-4)}{y}$ _____

2 Simplify:

a $5x^2 - 2x - 8x - 9x^2$ _____

b $\dfrac{3ab^2}{-2b}$ _____

c $3ab^2 \times (-2b)$ _____

3 Write an algebraic expression for:

a the number of seconds in x minutes

_____

b the number of times 3 divides into r

_____

c the perimeter of this rectangle

PART D: NUMERACY AND LITERACY

1 What is the meaning of quotient?

_____

2 Write a simplified algebraic expression for the average of $x + 4$ and $3x$.

_____

3 The area of a trapezium is $A = \dfrac{1}{2}(a+b)h$ where a and b are the lengths of the parallel sides and h is the perpendicular height. Use the formula to find the area of this trapezium.

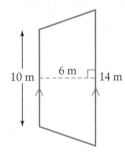

_____

_____

4 What word means to replace a variable with a number in an algebraic expression?

_____

5 **a** Write the 3 consecutive numbers before 5.

_____

b Write the 3 consecutive numbers before x.

_____

6 Complete: Only _____ terms can be added and subtracted.

7 Write an algebraic expression that has the value -2 when $c = 14$.

_____

9780170454537

① ALGEBRA 2

Name:

Due date:

Parent's signature:

THESE HOMEWORK TASKS COVER THE CURRENT TOPIC (ALGEBRA) AS WELL AS MIXED REVIEW. PART A SAYS 'NO CALCULATORS', PART B REVISES LAST WEEK'S WORK, PART C IS PRACTICE.

Part A	/ 8 marks
Part B	/ 8 marks
Part C	/ 8 marks
Part D	/ 8 marks
Total	/ 32 marks

PART A: MENTAL MATHS

🚫 Calculators not allowed

1 Evaluate:

a $11.95 + $23.90

_____

b $-6 + (-9)$ _____

c 25% of $120 _____

d $\dfrac{9}{10} + \dfrac{4}{5}$

_____

2 How many axes of symmetry has a rectangle?

_____

3 Find the range of 14, 20, 11, 19, 15, 12.

_____

4 Write these decimals in ascending order:

9.95, 9.909, 9.91, 9.9

_____

5 Complete this number pattern:

1, 3, 9, _____, 81, _____.

PART B: REVIEW

1 Write an algebraic expression for:

a the product of x and -9 _____

b the change from $60 after buying N hats at $7.40 each

_____

2 (2 marks) Complete this table for $y = \dfrac{1}{2}x$.

x	0	1	2	3	4
y					

3 Simplify:

a $-a - b - 2a + 9b$ _____

b $\dfrac{-12xy}{4yz}$ _____

c $-3r \times (-7r) \times 5$ _____

4 If $L = 5$ and $B = 8$, then find P if $P = 2(L + B)$.

_____

9780170454537

PART C: PRACTICE

› › Expanding and factorising expressions

1 List 3 factors of $5xy^2$.

_____

2 Expand:

a $3(2a - 7)$ _____

b $-6v(7vw - 2v)$

_____

3 Factorise $p + 3p^2$. _____

4 Find the highest common factor of 12 and 24.

_____

5 Expand and simplify:

a $8(b + 2) - 3b + 1$

_____

_____

b $x(2y - 3) + y(x - 4)$

_____

_____

6 Factorise $27mn^2 - 6m^2n$.

_____

PART D: NUMERACY AND LITERACY

1 Why is $2n$ a factor of $10n^2$?

_____

_____

2 Write a simplified algebraic expression for the area of this triangle.

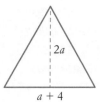

_____

_____

3 Find the highest common factor of $30xyz$ and $40xy^2$.

_____

4 **a** What word means to remove the brackets in an algebraic expression?

_____

b What is the opposite of this word?

_____

5 Write a simplified algebraic expression for the perimeter of this rectangle.

_____

_____

6 (2 marks) Evaluate 12×98 by expanding $12 \times (100 - 2)$. Show working.

① ALGEBRA 3

> PART D QUESTIONS ON THE NEXT PAGE CAN BE TRICKY BECAUSE THEY ASK YOU TO WRITE ABOUT YOUR MATHS, USING THE RIGHT TERMINOLOGY.

HOMEWORK

PART A: MENTAL MATHS

🖩 Calculators not allowed

1 Find x.

2 Find the mode of 1, 4, 9, 8, 3, 1, 10, 3.

3 Write the formula for the area of a circle with radius r.

4 Evaluate:

a $(-8)^2$ _____

b $524 - 380$

c $\sqrt[3]{27}$ _____

d 1% of $70

5 Write Pythagoras' theorem for this triangle.

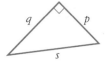

PART B: REVIEW

1 True or false? $8 - 3(a - 3) = -3a - 1$.

_____

2 Complete each factorisation.

a $n^2 + mn^2 =$ _____ $(1 + m)$

b $18y - 6xy =$ _____ $(3 - x)$

c $28f - 7f^2 + 35fg = 7f($_____$)$

3 Expand and simplify:

a $-2(8 - 5x)$ _____

b $a(3a - 6) + a(a - 7)$

_____

_____

4 Factorise:

a $10y - 3y^2$ _____

b $\dfrac{5}{6}st - \dfrac{5}{6}fs$ _____

PART C: PRACTICE

› Expanding binomial products
› Algebra revision

1 Simplify:

a $18k^2 \div 3k \times 2$ _____

b $8a - 2a^2 + 15a - a^2$ _____

c $\dfrac{5mn^3}{10n}$ _____

2 Write an algebraic expression for the area of this triangle.

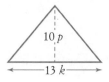

3 Expand and simplify:

a $(a + 2)(a + 3)$

_____

_____

_____

b $(2y + 3)(y - 6)$

_____

_____

_____

c $(5p - 1)(10 - 4p)$

_____

_____

_____

4 A rhombus has sides of length $y - 2$. Write an algebraic expression for its perimeter.

_____

PART D: NUMERACY AND LITERACY

1 What is the answer to a multiplication called?

_____

2 Write an algebraic expression for the number that is 8 more than half of m.

_____

3 Check that $-3a - 6 = -3(a + 2)$ by substituting a value of a into both sides of the equation.

_____

_____

4 (2 marks) **a** What does binomial mean?

_____

_____

b Give an example of a binomial product.

_____

5 In an algebraic expression such as $4n + 9$, what is the 'n' called?

_____

6 In a factory there are x managers each earning $90\ 500 annually and y workers each earning $62\ 000 annually. Write an expression for the total of the salaries the factory pays.

_____

7 What does **HCF** stand for?

_____

_____

HOMEWORK

HW

1 SUBSTITUTION PUZZLE

21	29

10

28	25	11	19	8	6	14	25	10	12

21	17

17	2	6	9	30	1	21	19	22

20	1	10	30

17	30	10	19	4	17

21	27

28	12	10	23	9

2	29

10

19	8	6	14	3	10	12

,

20	1	14	19

21	17

10

28	3	2	30	3	10	23	20	11	25

17	11	6	14	20	1	21	19	22

30	1	10	30

17	20	10	27	4	17

21	19

16	12	10	23	14

11	29

10

29	10	25	6	21	19	22

?

24	14	1	21	23	12	9

Evaluate the 30 expressions below, using the given values, and match each answer to a capital letter displayed in the 'Key' below. Then use the question number and the answer letter to decode the above message.

Given values

$a = 8.4$ $b = 0$ $c = 10$ $d = 4$ $e = 24$ $h = 7$ $k = -16$ $n = 6$

$p = 1$ $r = -1$ $s = 14$ $t = 0.5$ $w = -4$ $x = -10$ $y = 12$

1 $3p + 4$ ____   **2** $4d - 2$ ____   **3** $h^2 - 10$ ____   **4** $7 - c$ ____   **5** $2n^2$ ____

6 $\dfrac{-k}{4}$ ____   **7** $11 - 4t$ ____   **8** $(y - 5)^2$ ____   **9** $\sqrt{6e}$ ____   **10** $3x^2$ ____

11 $\dfrac{5a}{7}$ ____   **12** $s^2 - 18$ ____   **13** $2kr$ ____   **14** $\dfrac{3b}{10}$ ____   **15** $d^2 + e^2$ ____

16 $10 - 5w$ ____   **17** $\sqrt{hy - 20}$ ____   **18** $3(w + r)$ ____   **19** $\dfrac{-2c}{n - 1}$ ____   **20** $2s - 10p$ ____

21 $\dfrac{4b + 5}{b - 5}$ ____   **22** $-2ax$ ____   **23** $\dfrac{3d}{t}$ ____   **24** $5(n - r)$ ____   **25** $2w^2 - 4$ ____

26 $(k + e)^2$ ____   **27** $\dfrac{1}{2}st$ ____   **28** $\dfrac{(p + y)h}{7}$ ____   **29** ac^2 ____   **30** $20 - 3b$ ____

Key

A	300	F	840	M	4	P	13	U	49
B	64	G	168	N	3.5	R	28	V	35
C	24	H	7	N	−4	R	39	W	72
D	−3	I	−1	O	14	S	8	X	32
E	12	K	9	O	6	T	18	Y	−15
E	0	L	178	P	30	T	20	Z	592

STARTUP ASSIGNMENT 2 ②

THIS ASSIGNMENT COVERS BASIC SKILLS AND THINGS YOU SHOULD KNOW TO LEARN PYTHAGORAS' THEOREM PROPERLY.

PART A: BASIC SKILLS   / 15 marks

1 Simplify $\sqrt{\dfrac{50}{98}}$. _____

2 Whiteboard markers are 3 for $6.95 or 5 for $11.95. Which is the better buy?

_____

3 Factorise:

  a $15a - 35$ _____

  b $12xy + 14x - 8y$ _____

4 Complete: 1 m^2 = _____ cm^2

5 Write $\dfrac{1}{8}$ as a percentage. _____

6

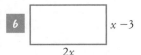

For this rectangle, find:

  a the perimeter _____

  b the area. _____

7 Evaluate $4 \div \dfrac{1}{3}$. _____

8 Draw an obtuse-angled triangle.

9 Simplify $m^6 \div m^2$. _____

10 Of the 75 teachers at Southvale High School, 32% are male. How many is this?

_____

11 Simplify $3m + 4mn - 2m + 6mn$.

_____

12 Simplify 36 : 20. _____

13 Find the value of x in this diagram.

   _____

PART B: SQUARE ROOTS AND TRIANGLES   / 25 marks

14 List the first 6 square numbers.

_____

15 How many degrees in 3 right angles?

_____

16 Draw a scalene right-angled triangle.

17 Draw an isosceles right-angled triangle.

18 What are the sizes of the 3 angles in the triangle you drew for question **17**?

_____

19 Find the value of x in each diagram below.

a

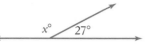

$x =$ _____

b

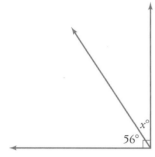

$x =$ _____

20 Evaluate:

a 20^2 _____

b 2.7^2 _____

c $0.4^2 + 1.6^2$ _____

d $\sqrt{2.25}$ _____

e $\sqrt{10^2 + 24^2}$ _____

21 Name the hypotenuse of each triangle.

a

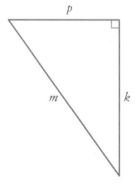

_____

b

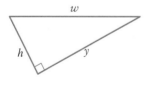

_____

22 Find the value of r in this triangle.

_____

23 True or false?

a $\sqrt{16+4} = \sqrt{16} + \sqrt{4}$ _____

b $\sqrt{16 \times 4} = \sqrt{16} \times \sqrt{4}$ _____

c $\sqrt{16-4} = \sqrt{16} - \sqrt{4}$ _____

24 Evaluate to 2 decimal places:

a $\sqrt{24}$ _____

b $\sqrt{11^2 + 5^2}$ _____

c $\sqrt{12^2 - 6^2}$ _____

d $\sqrt{10^2 - 3.3^2}$ _____

25 True or false?

a $9^2 + 12^2 = 15^2$ _____

b $8^2 + 10^2 = 14^2$ _____

c $87^2 - 60^2 = 63^2$ _____

PART C: CHALLENGE Bonus / 3 marks

Remove 4 matches to leave 4 triangles.

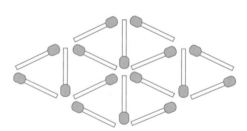

A PAGE OF RIGHT-ANGLED TRIANGLES ②

THIS WORKSHEET CAN BE USED IN DIFFERENT WAYS,
DEPENDING ON WHAT YOUR TEACHER DECIDES.
IT WILL HELP YOU LEARN ABOUT PYTHAGORAS' THEOREM.

Teacher's tickbox

For each triangle:

❏ circle the hypotenuse

❏ measure the length of each unknown side
correct to the nearest 0.5 cm

❏ square the length of each side

❏ write Pythagoras' theorem

❏ show Pythagoras' theorem is true

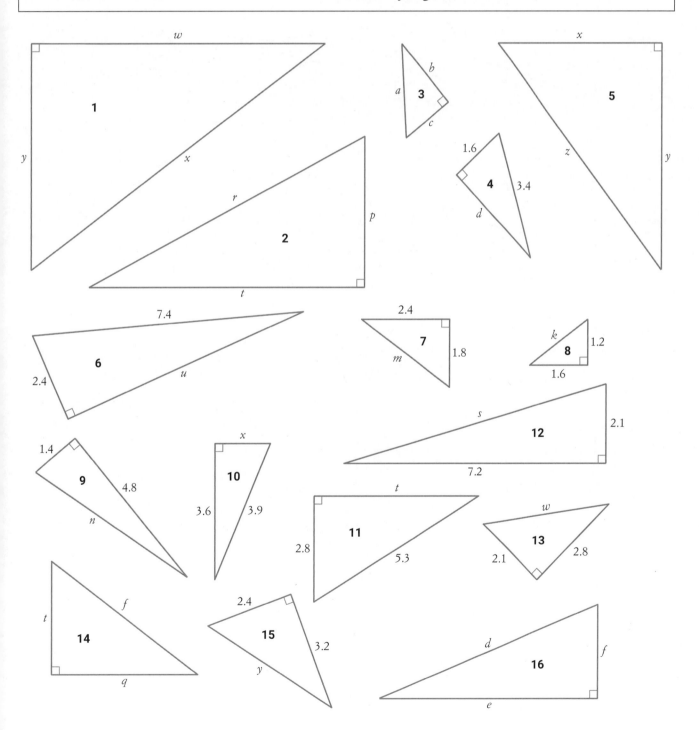

Chapter 2 Pythagoras' theorem **17**

ARE YOU MASTERING PYTHAGORAS' THEOREM YET?
DON'T WORRY ... IT TAKES TIME TO LEARN A NEW SKILL.
THERE'S MIXED ANSWERS AT THE BOTTOM OF THE PAGE.

Find the value of each variable, correct to 2 decimal places where appropriate.

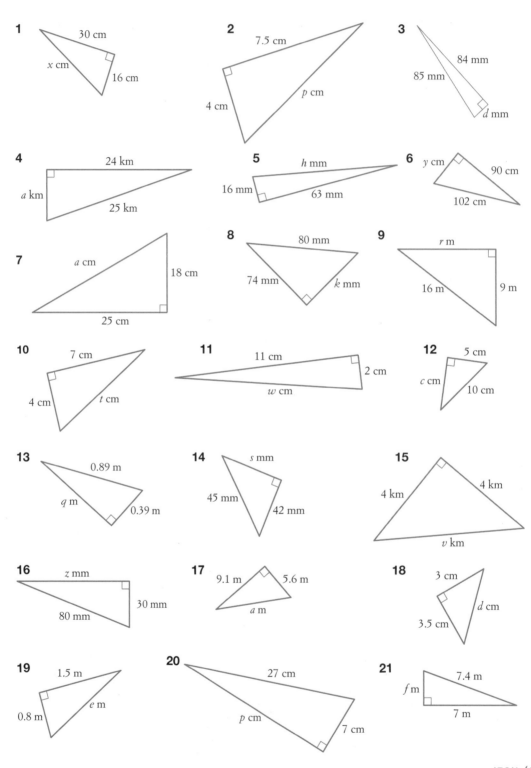

Mixed answers: 0.8, 1.7, 34, 74.16, 8.06, 8.66, 7.17, 11.18, 7, 30.40, 48, 2.4, 8.5, 13, 65, 13.23, 30.81, 5.66, 26.08, 10.69, 16.16, 4.61.

PYTHAGORAS' THEOREM FIND-A-WORD ② 2

ALL THE WORDS FROM THIS TOPIC ARE IN THERE SOMEWHERE.
TRY CHOOSING AN UNUSUAL LETTER FROM EACH WORD FIRST.

```
Y V O B T Q T O J Y Q Z K F O O R P M Y U P I U X
X Q E R A U Q S R U O F G A W C J M E R O E H T A
Q C X T H R E E Z T S E T M T O O R R P N T S U K
Y L P P A U Z A E R A J C I D E D N U O R O N U Y
Y Q P S L A M I C E D L A Y R K D I A G O N A L K
I P E V C P I V C Y A D T C A X E Y E C S X R R R
S L R E V I F U P N K X D X E T A M I X O R P P A
Y K I R K E Z X O D P P Y T H A G O R A S O M E T
Y J M U G D D I B R E V O C S I D F J O E D I S R
P N E Q C N T X E K E M C E Y M H P U O Z H Y B B
T U T J I A R Y N T I G K G C T A J N Y X A B H X
G M E R R E Z J U S L X T G K J I N Z I V J M M G
Y B R R T T M T P K P J L T E U M J O L R I H D T
B E I R D Z I S B C X L C W L E K Y J B X T D A S
U R O U N T C M H T G N E L J Q W R G D C D T I E
I H Q L S E L G N A I R T D E L G N A T H G I R G
S L P B D R Z E S U N E T O P Y H S U R D L W T N
M P U J H V I N A L U M R O F G Y C O Q Z B G F O
U S E T I S O P P O W U E S R E V N O C J W T S L
B V E H Y J N A P Y B J H D E S Y F W Y S G U F C
```

Find these words in the puzzle above. They appear across, up and down, and diagonal, and can be backwards as well as forwards.

APPLY	APPROXIMATE	AREA	CONVERSE
DECIMAL	DIAGONAL	DISCOVER	EXACT
FIVE	FORMULA	FOUR	HYPOTENUSE
IRRATIONAL	LENGTH	LONGEST	NUMBER
OPPOSITE	PERIMETER	PROOF	PYTHAGORAS
RIGHT-ANGLED	ROOT	ROUNDED	SHORTER
SIDE	SQUARE	SUBSTITUTE	SURD
TEST	THEOREM	THREE	TRIAD
TRIANGLE			

Chapter 2 Pythagoras' theorem

② PYTHAGORAS' THEOREM 1

NOW LET'S PRACTISE AND REVISE PYTHAGORAS' THEOREM. WHAT'S THE LONGEST SIDE OF A RIGHT-ANGLED TRIANGLE CALLED AGAIN?

Name:

Due date:

Parent's signature:

Part A	/ 8 marks
Part B	/ 8 marks
Part C	/ 8 marks
Part D	/ 8 marks
Total	/ 32 marks

PART A: MENTAL MATHS

🚫 Calculators not allowed

1 Write 21:08 in 12-hour time. _____

2 Complete: 60.8 cm = _____ mm

3 Find the area of this trapezium.

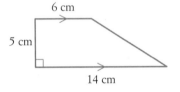

6 cm
5 cm
14 cm

_____

4 Evaluate 44×5. _____

5 Factorise $8m - 4m^2$. _____

6 Find the perimeter of this shape.

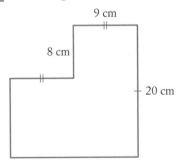

9 cm
8 cm
20 cm

_____
_____

7 Evaluate $230 \div 5$. _____

8 Evaluate $\dfrac{2}{7} - \dfrac{3}{14}$.

_____

PART B: REVIEW

1 Complete: 4120 kg = _____ t

2 Evaluate $8^2 + 3^2$. _____

3 Evaluate correct to 2 decimal places:

  a $\sqrt{95}$ _____

  b $\sqrt{15^2 - 10^2}$ _____

4 Solve $x - 30 = 145$. _____

5 Find p.

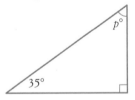

$p°$
$35°$

_____

6 Solve $2y + 18 = 24$.

_____

7 Solve $\dfrac{m}{12} = 3$. _____

PART C: PRACTICE

📝 › Pythagoras' theorem
› Finding the hypotenuse
› Finding a shorter side

1 **a** Which side of this triangle is the hypotenuse?

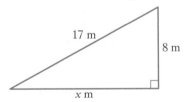

17 m
8 m
x m

_____

b Find x.

_____

2 Find k as a surd.

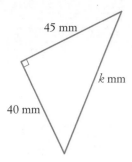

45 mm

k mm

40 mm

_____

3 Find p correct to one decimal place.

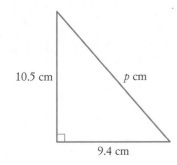

10.5 cm

p cm

9.4 cm

_____

4 Circle the surds from this list of square roots: $\sqrt{9}$, $\sqrt{25}$, $\sqrt{49}$, $\sqrt{50}$, $\sqrt{33}$, $\sqrt{69}$.

5 Find y correct to 2 decimal places.

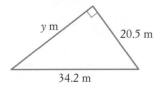

y m

20.5 m

34.2 m

_____

6 (2 marks) Construct a right-angled triangle with perpendicular sides of length 25 mm and 60 mm and by measurement find the length of its hypotenuse.

PART D: NUMERACY AND LITERACY

1 Describe the meaning of each word:

a perimeter

_____

_____

b hypotenuse

_____

_____

2 Write Pythagoras' theorem for this triangle.

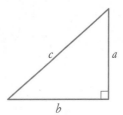

c

a

b

_____

3 What name is given to a square root that cannot be expressed as an exact decimal?

_____

4 Write 3^2 in words.

_____

5 a Draw a right-angled isosceles triangle.

b Write the size of each angle in the above triangle.

_____

6 Complete: In a right-angled triangle, the square of the hypotenuse is equal to

_____

_____

② PYTHAGORAS' THEOREM 2

HOW'S YOUR UNDERSTANDING OF PYTHAGORAS' THEOREM NOW? KEEP PRACTISING, YOU'LL GET THERE!

Name:

Due date:

Parent's signature:

Part A	/ 8 marks
Part B	/ 8 marks
Part C	/ 8 marks
Part D	/ 8 marks
Total	/ 32 marks

PART A: MENTAL MATHS

🚫 Calculators not allowed

1 Evaluate 39×11. _____

2 Evaluate $\dfrac{2}{3} - \dfrac{1}{5}$.

_____

3 Factorise $6xy^2 + 2y$. _____

4 Expand $-2(x - 4)$. _____

5 Find $\dfrac{2}{3}$ of $36. _____

6 Find the mean of 1, 2, 8, 3, 6.

_____

7 Kathy pays a grocery bill of $83.45 with a $100 note. Calculate the change.

_____

8 Use 3 letters to name the marked angle.

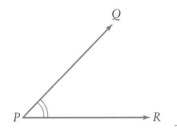

_____

PART B: REVIEW

1 Find the value of each variable.

a

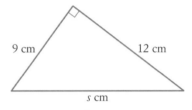

9 cm   12 cm
s cm

_____

b

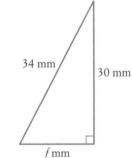

34 mm   30 mm

f mm

_____

2 Select the surds from this list of square roots: $\sqrt{75}$, $\sqrt{12^2}$, $\sqrt{41}$, $\sqrt{16}$, $\sqrt{28}$, $\sqrt{100}$.

_____

3 Write Pythagoras' theorem for each triangle.

a

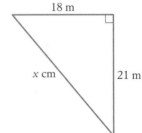

18 m

x cm   21 m

_____

b

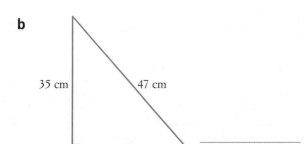

35 cm · 47 cm · p cm

4 Find, correct to one decimal place, the value of each variable in Question **3**.

a _____ **b** _____

5 Find k as a surd.

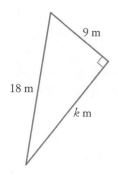

9 m · 18 m · k m

_____

PART C: PRACTICE

1 **a** Show that (8, 15, 17) is a Pythagorean triad.

_____

_____

b Create another Pythagorean triad by multiplying every number in the triad above by the same number.

_____

_____

2 Test whether this triangle is right-angled.

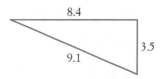

8.4 · 9.1 · 3.5

_____

3 Find the perimeter of this triangle.

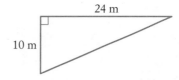

24 m · 10 m

_____

4 Find, correct to one decimal place, the area of this rectangle.

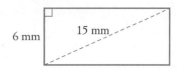

6 mm · 15 mm

_____

5 Show that this triangle is right-angled and mark the right angle on it.

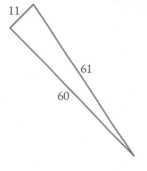

11 · 61 · 60

_____

6 This rectangle has a length of 14 mm and a diagonal of 16 mm. What is its height, d, correct to 2 decimal places?

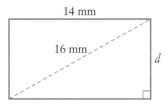

_____

_____

7 Find x as a surd.

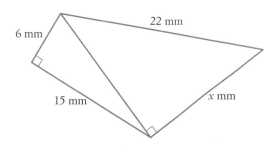

_____

_____

PART D: NUMERACY AND LITERACY

1 What does theorem mean?

_____

_____

2 What does Pythagoras mean?

_____

_____

3 The foot of a ladder is placed 0.9 m from the base of a wall. If the ladder reaches 2.1 m up the wall, how long is the ladder, correct to one decimal place?

_____

_____

4 Find, correct to 2 decimal places, the length of this playground slide.

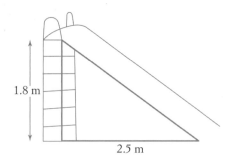

_____

_____

5 Describe what a surd is.

_____

_____

_____

6 Complete: The hypotenuse is _____ the right angle.

7 A ship sails 5.2 km west, then 6.1 km as shown until it is due north of its starting point. How far is it from its starting point, correct to 2 decimal places?

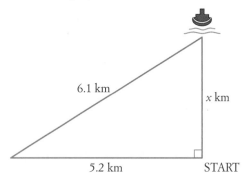

_____

_____

8 How long is the diagonal in a square of length 4 cm? Write your answer as a surd.

_____

_____

HAVE YOU WORKED OUT WHY SOME QUESTIONS ARE BLUE AND SOME ARE GREEN? THERE'S EVEN A RED ONE NEXT PAGE. YOU'LL NEED MORE TIME TO DO THAT ONE.

Name:

Due date:

Parent's signature:

Part A	/ 8 marks
Part B	/ 8 marks
Part C	/ 8 marks
Part D	/ 8 marks
Total	/ 32 marks

PART A: MENTAL MATHS

🚫 Calculators not allowed

1 Convert 45% to a simple fraction.

_____

2 Convert 5.8% to a decimal.

_____

3 Evaluate $13.26 \div 6$.

_____

4 Find the average of 10 and -8.

_____

5 Round $235.2685 to the nearest cent.

_____

6 Find the area of this triangle.

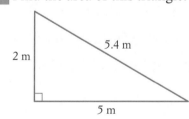

_____

7 Simplify $10 - 8x \div x$.

_____

8 If $a = -2$, then evaluate $14 - 3a$.

_____

PART B: REVIEW

1 Write Pythagoras' theorem for each triangle.

a

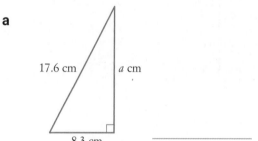

_____

b

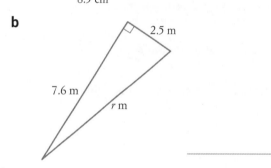

_____

2 Find, correct to 2 decimal places, the value of each variable in question **1**.

a _____

b _____

3 Test whether (3.5, 8.4, 9.1) is a Pythagorean triad.

_____

_____

4 The formula $\left[x, \frac{1}{2}(x^2 - 1), \frac{1}{2}(x^2 + 1)\right]$ gives a Pythagorean triad. Substitute $x = 5$ to find a Pythagorean triad.

_____

_____

HOMEWORK

HW

5 Find as a surd the value of each variable.

a

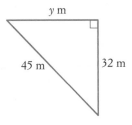

_y_ m

45 m   32 m

_____

b

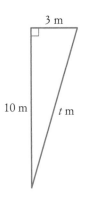

3 m

10 m   _t_ m

_____

PART C: **PRACTICE**

Pythagoras' theorem revision

1 Test whether each triangle is right-angled.

a

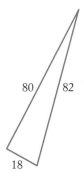

80   82

18

_____

b

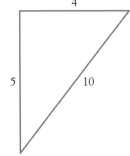

4

5   10

_____

2 Find the perimeter of each shape, correct to 2 decimal places where necessary.

a

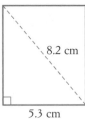

8.2 cm

5.3 cm

_____

_____

b

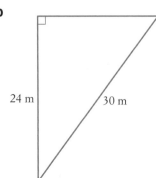

24 m   30 m

_____

_____

3 Find the area of each shape in question **2**, correct to 2 decimal places where necessary.

a _____

b _____

4 (2 marks) Plot the points _A_(2, 3) and _B_(4, 7) on a number plane and use Pythagoras' theorem to find the length of interval _AB_, correct to one decimal place.

_____

_____

PART D: NUMERACY AND LITERACY

1 What is the longest side of a right-angled triangle called?

_____

2 A ladder 3 m long leans against a wall. The foot of the ladder is 0.8 m from the base of the wall. Calculate how far up the wall the ladder reaches, correct to one decimal place.

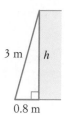

_____

3 Find, correct to 2 decimal places, the length of this ramp.

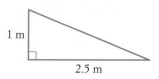

4 State Pythagoras' theorem in words.

_____

_____

_____

_____

5 The size of a TV screen is described by the length of its diagonal. Find, correct to one decimal place, the height of a TV screen with a 109 cm diagonal if its length is 95 cm.

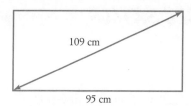

_____

6 A ship sails 30 km south and then 35 km east. How far is it from its starting point, correct to 2 decimal places?

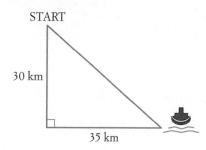

_____

7 Describe what a Pythagorean triad is.

_____

_____

_____

8 A square has a diagonal of length 8 cm. Find the length of each side of the square as a surd.

_____

_____

(3) STARTUP ASSIGNMENT 3

IN THIS TOPIC, WE'LL BE REVISING FRACTIONS, DECIMALS, PERCENTAGES, RATIOS AND RATES.

PART A: BASIC SKILLS /15 marks

1 Simplify $14:10$. _____

2 What is the angle sum of a quadrilateral?

_____

3 Find the value of x in this diagram, correct to 2 decimal places.

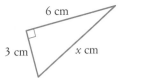

6 cm

3 cm x cm   _____

4 For the triangle above, calculate:

a its area _____

b its perimeter. _____

5 Write 2 negative numbers that have a sum of -5. _____

6 Expand $-3(a + 8)$. _____

7 Find the average of 6, 9, and 15. _____

8 Find the median of these values:

16, 17, 11, 12, 14, 25, 10 _____

9 If $r = 1$, evaluate $r - 9$. _____

10 Simplify $-3m \times 4m$. _____

11 Write the factors of 6. _____

12 When rolling a die, what is the probability that the number that shows up is a factor of 6? _____

13 Evaluate: $\dfrac{3}{4} + \dfrac{1}{5}$. _____

14 What fraction of this rectangle is shaded?

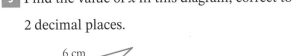

_____

PART B: NUMBER /25 marks

15 Convert each number to a decimal.

a $\dfrac{3}{4}$ _____   **b** $\dfrac{2}{3}$ _____

c $\dfrac{9}{100}$ _____   **d** 16% _____

16 Simplify:

a $\dfrac{16}{40}$ _____   **b** $\dfrac{35}{100}$ _____

17 Convert each number to a simple fraction.

a 0.4 _____   **b** 0.04 _____

c 0.125 _____   **d** 70% _____

18 Round 38.765 21 to:

a the nearest whole number _____

b 2 decimal places. _____

19 Evaluate:

a $8.4 \div 100$ _____

b 0.129×100 _____

20 Evaluate:

a $\dfrac{3}{5} \times 70$ _____   **b** $\dfrac{3}{8} \times 100$ _____

21 Convert each number to a percentage.

a $\dfrac{2}{5}$ _____   b $\dfrac{1}{3}$ _____

c 0.5 _____   d 0.0375 _____

22 Find 50% of $17. _____

23 Complete: $2 : 5 =$ _____ $: 60$

24 Divide $450 in the ratio $2 : 7$.

_____

_____

_____

_____

25 What is the cost of $\dfrac{1}{2}$ kg of tomatoes at

$3.80 per kilogram? _____

26 Find the cost (to the nearest cent) of 33.24 L of

petrol at 143.9c/L. _____

PART C: CHALLENGE   Bonus   / 3 marks

A group of 150 Year 9 students wanted to send the same message to all the students in the group.

The first student messaged 3 other students, each of whom messaged another 3, and so on, until all 150 students had been contacted. If nobody in the group received the same message twice, how many students did not need to send on the message?

_____

(3) DISCOUNTS AND SPECIAL OFFERS

HERE ARE SOME REAL-LIFE PROBLEMS ON SPENDING AND SAVING MONEY. THIS WILL HELP YOU BECOME A SMART CONSUMER.

Round all price answers to the nearest cent and all percentages to one decimal place.

1 A box of 25 Cheapo pens costs $20.95 while a 10-pack of the same brand costs $7.95.
Which is the better buy?

2 Marty's Sports Warehouse is having a '15% off' sale on all goods. Calculate the sale prices if these are the original prices:

a **Roller blades** were $165

b **Netball skirts** were $52

c **Soccer balls** were $78

d **Home gym** was $480

3 Hannah bought a $1870 computer for $1450. What was the discount as a percentage of the original price?

4 Jeans that cost $45 are discounted by 14%. What is their sale price?

5 A microwave oven is on sale at 3 stores:
Good Gals: $244 less 5% discount
Hardly Normal: $259 less $12\frac{1}{2}$% discount
J-Mart: 4 monthly payments of $55.20
Which store gives the best offer?

6 For each item, find the percentage discount on the original price:

a TV set: was $348 now $299

b Headsets: was $25 now $23

c Toy truck: was $81 now $64

d Basketball backboard: was $101 now $84

7 Sunglasses were selling at the discounted price of $18.36. If the discount was $4.59, find:

a the original price

b the percentage discount.

8 Goo glue sticks come in 3 sizes:
Small: 8 g for $1.40
Medium: 21 g for $2.57
Large: 35 g for $3.59

a Which size is best value for money?

b Why do you think that this size is the most economical?

9 At Pizza World, pizzas cost $9.50, a bottle of cola costs $3.45 and garlic bread is $3.10.
In which of the following deals do you save more money?
Hungry Pack ($23.90): 2 pizzas, 1 cola, 2 garlic breads.
Triple Treat ($29.90): 3 pizzas, 2 colas.

9780170454537

10 Joggers were selling for $75 a pair. If their original price was $92, calculate the discount as a percentage of the original price.

11 Big V are having a '12% off' sale. Calculate the *original* marked prices of each item:

  a game system, sale price $364.32

  b tablet device, sale price $492.80

  c racing bike, sale price $247.28

  d outdoor table, sale price $83.60

12 Two travel agents offer trips to Fiji. Deadset Tours offers 11 nights for $780. Balmy World Travel offers 13 nights for $910.

  a Which is better value for money?

  b What would Deadset charge for 13 nights based on their 11-night rate?

13 Jane bought a home cinema priced at $1896 from 3D World. She paid $\frac{1}{3}$ deposit followed by monthly payments of $60.90 over 2 years. Calculate:

  a the deposit paid

  b the total amount paid

  c the extra amount paid (interest)

  d the interest as a percentage of the original price

14 Souraya bought theme park tickets for $36.50, a saving of $8.40 off their original price. What was the percentage discount?

15 A tool set with a marked price of $90 is sold with a discount of 7.5%. What was the sale price?

16 Snappy Photo Shop prints 36 photos for $4.95 while Prints Charming charges $3.95 for 25 photos. Which store gives the better value?

17 Paul bought a BBQ for $215, saving $69.

  a What was the original price?

  b What percentage discount did he receive on this price?

18 Regan saved $36 when he bought a lawn mower for $369. What was the percentage discount?

19 Find the sale price of a mobile phone priced at $189 discounted by 15%.

20 Which pack offers better value for money?

  Feasty Pack: 10 chicken pieces for $21.95

  Family Pack: 16 chicken pieces for $32.95

(3) TIME CALCULATIONS

SOME PEOPLE FIND 24-HOUR TIME AND TIME CALCULATIONS TRICKY BECAUSE THERE ARE 60 MINUTES IN AN HOUR, NOT 100. LEARN HOW TO DO THESE CORRECTLY.

1 Convert each time to 24-hour time.

  a 6:30 a.m. _____

  b 4:20 p.m. _____

  c 11:05 p.m. _____

  d 2:45 p.m. _____

  e 7:18 p.m. _____

  f 10:56 a.m. _____

2 Convert each time to 12-hour time.

  a 07:05 _____

  b 18:55 _____

  c 12:30 _____

  d 20:30 _____

  e 03:44 _____

  f 15:17 _____

3 What is the time:

  a 6 hours after 3 p.m.? _____

  b 5 hours after 8 a.m.? _____

  c 14 hours after 17:00? _____

  d 7 hours before 10 p.m.? _____

  e 10 hours before 19:00? _____

  f 16 hours before 12 midnight? _____

  g 13 hours after 9:30 a.m.? _____

  h 9 hours before 14:40? _____

  i $8\frac{1}{2}$ hours after 6 p.m.? _____

4 Use the [° ' "] or [D°M'S] key on your calculator to convert each time to hours and minutes.

  a 9.5 hours _____

  b 4.3 hours _____

  c 12.6 hours _____

  d 7.4 hours _____

  e 15.7 hours _____

  f 10.8 hours _____

  g 5.25 hours _____

  h 8.3 hours _____

5 Convert each time to hours and minutes.

  a 378 minutes _____

  b 236 minutes _____

  c 173 minutes _____

  d 685 minutes _____

  e 427 minutes _____

  f 404 minutes _____

9780170454537

6 What is the time:

a 6 hours 20 minutes after 10:15 p.m.? _____

b 5 hours 24 minutes before 21:30? _____

c 10 hours 4 minutes after 12:10 p.m.? _____

d 3 hours 35 minutes before midday? _____

7 How many hours and minutes between:

a 10 a.m. and 6 p.m.? _____

b 8:35 a.m. and 5:15 p.m.? _____

c 2:40 p.m. and 8 p.m.? _____

d 5:40 a.m. and 1:20 p.m.? _____

e 15:11 and 20:24? _____

f 11:37 and 22:20? _____

8 What is the time:

a 233 minutes after 11:47 p.m.? _____

b 109 minutes before 2:30 p.m.? _____

c 85 minutes after 7:54 a.m.? _____

d 174 minutes before 1:00 p.m.? _____

e 386 minutes after 18:40? _____

f 270 minutes before 15:51? _____

9 This map shows the time differences between the Australian states relative to AEST.

Australian Western Standard Time (AWST)

Australian Central Standard Time (ACST)

Australian Eastern Standard Time (AEST)

−2 hours
8:00 a.m.

−$\frac{1}{2}$ hour
9:30 a.m.

Zero
10:00 a.m.

When it is 3:30 p.m. in Sydney, what is the time in:

a Hobart? _____

b Alice Springs (NT)? _____

c Perth? _____

d Brisbane? _____

e Melbourne? _____

f Adelaide? _____

10 Use the map above to help you find the time in Sydney when it is:

a 4 a.m. in Canberra _____

b 6:30 p.m. in Perth _____

c 1:15 p.m. in Darwin _____

d 10:47 a.m. in Cairns (Qld) _____

e 15:05 in Adelaide _____

f 07:20 in Hobart _____

EVEN THOUGH THIS TOPIC IS ABOUT NUMBERS, THERE ARE A LOT OF WORDS IN IT TO LEARN AS WELL. THIS CROSSWORD IS NOT EASY TO SOLVE, BUT DON'T GIVE UP!

Clues across

1 $0.1 =$ _____ %

3 A number that uses a point to separate the whole from the part

4 The top number in a fraction

6 This is made when the selling price is more than the cost price

8 To convert a fraction to a percentage, multiply by this

11 The second place after the decimal point

12 A 'Q'-word meaning 'amount'

16 A fraction out of 100 with the symbol %

19 0.5 as a fraction

20 The price at which the item was bought by the store

21 $\dfrac{1}{100}$ of a dollar

23 A decimal that does not repeat

24 To determine roughly the value of something

25 To make simpler

28 A decimal that repeats

29 To change from one form to another

30 At a sale, the amount reduced from the original price

31 The price at which an item is sold is the _____ price

Clues down

1 $33\dfrac{1}{3}$% as a fraction is one-_____

2 Rounding to one decimal place is to the nearest _____

5 From smallest to largest

7 Any number than can be written in fraction form is called a r_____ number

9 To approximate a decimal to a given number of places

10 The bottom number in a fraction

13 $\dfrac{1}{4}$ as a percentage is _____ %

14 From largest to smallest is _____ order

15 To find the value of an expression

17 To make bigger

18 A number expressed in the form $\dfrac{a}{b}$

22 To make smaller

26 8.107 has 3 decimal _____

27 $\dfrac{1}{2} =$ _____ %

3 INTEGERS AND DECIMALS

HEY, I'M MITCH. SOME CLUES FOR THIS HOMEWORK ASSIGNMENT: 'ASCENDING' MEANS FROM SMALLEST TO LARGEST, 'TERMINATE' MEANS TO STOP

Name:

Due date:

Parent's signature:

Part A	/ 8 marks
Part B	/ 8 marks
Part C	/ 8 marks
Part D	/ 8 marks
Total	/ 32 marks

PART A: MENTAL MATHS

🔢 Calculators not allowed

1 Evaluate:

a $22 - 3^2 \times 2$ _____

b 80×9 _____

c $-12 + 5 + (-8)$ _____

2 Find x.

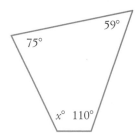

_____

3 List all of the factors of 6.

_____

4 Given that $15 \times 6 = 90$, evaluate 1.5×6.

_____

5 Simplify $-x + y - y + 2x$. _____

6 Find the median of this set of data.

```
            •
    •   •   •
•   •   •   •   •
1   2   3   4   5
```

_____

PART B: REVIEW

1 Convert 0.64 to:

a a percentage _____

b a simple fraction _____

2 Find 85% of $450. _____

3 Simplify $\frac{18}{45}$. _____

4 Evaluate:

a 18^3 _____

b $\frac{-462}{-11}$ _____

c $\sqrt{400 - 39}$ _____

5 Evaluate $\frac{2}{3} + \frac{1}{5}$.

_____

_____

9780170454537

PART C: PRACTICE

› Integers
› Decimals

1 Write these decimals in ascending order:

8.8, 8.89, 8.714.

_____

2 Evaluate:

a $-9 + [3 \times (-2)] \div 6.$

_____

b $4.017 + 18.24$

_____

_____

c $28.4 - 13.341$

_____

_____

d 0.02×1.1

_____

_____

e $2.4 \div 0.08$

_____

_____

3 Convert to a decimal:

a $\dfrac{17}{40}$

_____

b $\dfrac{5}{18}$

_____

PART D: NUMERACY AND LITERACY

1 Complete: When dividing a decimal by 100, move the decimal point _____ places to the

_____.

2 Biljana earns $845.45 for working 37 hours per week. What is her hourly pay rate?

_____

3 What number gives 30 when divided by 0.9?

_____

4 Samantha ran 100 m in 10.45 seconds while James ran it in 11.27 seconds. Who was faster and by how many seconds?

_____

_____

5 Rounding a decimal to the nearest tenth is the same as rounding to how many decimal places?

_____

6 If x is a negative number, is x^2 always positive, always negative or can it be either positive or negative?

_____

7 Kate bought 31.2 L of petrol at $1.32 per litre. How much did it cost?

_____

8 Describe a **terminating decimal**, giving an example.

_____

_____

3 FRACTIONS AND PERCENTAGES

THIS TOPIC IS ALL ABOUT NUMBERS, WHICH YOU'LL USE EVERY DAY. LEARN HOW TO WORK WITH NUMBERS WITH A CALCULATOR BUT ALSO WITHOUT! DON'T FORGET TO ESTIMATE THE ANSWER FIRST.

Name:

Due date:

Parent's signature:

Part A	/ 8 marks
Part B	/ 8 marks
Part C	/ 8 marks
Part D	/ 8 marks
Total	/ 32 marks

PART A: MENTAL MATHS

🚫 Calculators not allowed

1 Round 2.189 to 2 decimal places.

_____

2 Simplify:

a $8x + 4y + 7x - 2y$. _____

b $\dfrac{9x^2 y}{3y}$ _____

3 Find the reciprocal of $1\dfrac{1}{4}$.

_____

4 Write the formula for the area of a triangle with base length b and perpendicular height h.

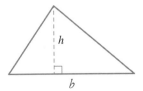

_____

5 Evaluate 75% of $28. _____

6 Find the mean of these values, correct to one decimal place.

12, 10, 9, 15, 17, 13.

_____

7 What is the probability of a person chosen at random having a birthday in a month beginning with J? _____

PART B: REVIEW

1 Convert $\dfrac{27}{7}$ to a mixed numeral. _____

2 Write the decimal shown by the arrow on each number line.

a

b

3 Evaluate:

a $\dfrac{-45}{15}$ _____

b $4^2 - 3 \times (-7)$ _____

4 Convert each number to a decimal and state whether they are recurring, terminating or neither.

a $\dfrac{4}{5}$

_____

b $\dfrac{5}{11}$

_____

c π

_____

9780170454537

PART C: PRACTICE

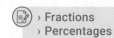
> Fractions
> Percentages

1 Convert:

a 45% to a simple fraction _____

b $\dfrac{74}{80}$ to a percentage _____

c $8\dfrac{3}{4}\%$ to a decimal _____

2 Evaluate:

a $4\dfrac{1}{5} - 2\dfrac{3}{8}$

_____

_____

b $\dfrac{3}{4} \div \dfrac{5}{12}$

_____

_____

3 Decrease $700 by 4.5%.

_____

4 Find $\dfrac{2}{5}$ of $65.

_____

5 Arrange these numbers in descending order:

$\dfrac{3}{8}$, 45%, 40.5%, $\dfrac{2}{5}$.

_____

PART D: NUMERACY AND LITERACY

1 Adrian scored 66 runs in a cricket match, which was 24% of his team's total. How many runs did his team score?

_____

_____

_____

2 Complete: Multiplying a number by 1.08 increases it by _____ %.

3 How much will Lauren pay for a $165 skirt that has been discounted by 15%?

_____

4 What percentage (correct to 2 decimal places) is:

a $510 of $7800?

_____

b 25 min of 2 hours?

_____

5 What is the answer when a fraction is multiplied by its reciprocal?

_____

6 Phillip's weekly pay increased from $754 to $835. Find:

a the increase in Phillip's pay.

_____

b the percentage increase (correct to one decimal place).

_____

3 PERCENTAGES, RATIOS AND RATES

I JUST NOTICED THAT THIS HOMEWORK ASSIGNMENT HAS A LOT OF PRACTICAL, REAL-LIFE MATHS. GET ALL THE PROBLEMS OUT AND YOU'LL BE A MATHS MASTER!

Name:

Due date:

Parent's signature:

Part A	/ 8 marks
Part B	/ 8 marks
Part C	/ 8 marks
Part D	/ 8 marks
Total	/ 32 marks

HW HOMEWORK

PART A: MENTAL MATHS

 Calculators not allowed

1 Convert $\dfrac{36}{8}$ to a mixed numeral. _____

2 Simplify $xy - px + 5xy + xp$. _____

3 a Write Pythagoras' theorem for this triangle.

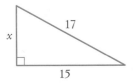

_____

b Find the value of x.

_____

_____

4 Evaluate $8.4 - 4.25$. _____

5 Arrange these integers in descending order:

$-5, 7, 0, 6, 3, -2$

_____

6 Find 15% of $40. _____

7 How many hours in one week? _____

PART B: REVIEW

1 Convert $\dfrac{7}{15}$ to a decimal. _____

2 Find $15\dfrac{1}{2}\%$ of 6 L in mL. _____

3 Write a decimal with 3 decimal places that can be rounded to 5.2.

_____

4 Evaluate:

a $\dfrac{8}{1000} + \dfrac{4}{10}$ _____

b $\dfrac{2}{5} \times \dfrac{25}{32}$ _____

c $(-2)^6$ _____

5 Increase $30.50 by 12%. _____

6 Write a fraction that is greater than $\dfrac{3}{5}$ but less than 65%.

_____

9780170454537

PART C: PRACTICE

> › Percentages
> › Ratios and rates
> › Time

1 Simplify 30 : 9. _____

2 Tom buys a refrigerator for $950 and sells it for $1200. Find:

   a the profit _____

   b the percentage profit (on the cost price) correct to one decimal place

   _____

3 How many hours and minutes between 3.30 p.m. and 10.45 p.m.?

   _____

4 A car travelled 276 km in 3 hours. What was its average speed?

   _____

   _____

5 Paula and Erik own a shop in the ratio 5 : 7. If the shop makes a profit of $18 900, what is Erik's share?

   _____

   _____

6 Find the simple interest earned on an investment of $5800 over 3 years at 4.5% p.a.

   _____

7 Katrina earned commission of $1207.50 for selling a car. If this was 3.5% of the price of the car, what is the price of the car?

   _____

PART D: NUMERACY AND LITERACY

1 Convert 108 km/h to m/s.

   _____

2 A film started at 2.45 p.m. and ran for 132 minutes.

   a Convert 132 minutes to hours and minutes.

   _____

   b At what time did the film finish?

   _____

3 Explain the difference between **profit** and **loss**.

   _____

   _____

   _____

   _____

4 A packet of bread rolls was discounted by 10% so that its sale price is $4.86. What was its original price?

   _____

5 Simplify the ratio 35 min : 2 h.

   _____

6 Bridie's heart beats 76 times per minute. How long will it take to beat 513 times?

   _____

7 Lyn invested $65 000 for 6 years and earned $35 100 in simple interest. What was the interest rate p.a.?

   _____

4 STARTUP ASSIGNMENT 4

WS WORKSHEET

PART A: BASIC SKILLS   / 15 marks

1 Simplify $14:10$. _____

2 Calculate 11.75% of $190 000. _____

3 Divide $510 in the ratio 11:6.
_____

4 Expand and simplify $m(3m - 4) + 2m(5 - m)$.
_____
_____

5 What fraction of 2 m is 8 cm? _____

6 Evaluate 10^5. _____

7 A rhombus is a square. True or false?
_____

8 A box of 1500 paper clips costs $29.50.
How much does one paper clip cost, to the
nearest cent? _____

9 Find the average of 6, 9 and 15. _____

10 Solve $\dfrac{2x - 5}{4} = 7$ _____

11 Find the value of k in the diagram below.

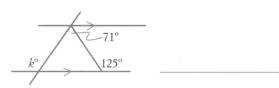
_____

12 How many sides has an octagon? _____

13 Complete: 2.1 L = _____ mL.

14 Find:

a the length of AB
in this triangle,
correct to 2 decimal
places _____

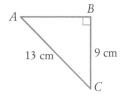

b the triangle's area. _____

PART B: TRIANGLE GEOMETRY

/ 25 marks

15 How many degrees in 4 right angles? _____

16 What is the angle sum of a triangle? _____

17 Evaluate:

a $4 \times \sqrt{24.01}$ _____

b $10 \times \sqrt{65.61}$ _____

c $\dfrac{63}{\sqrt{12.25}}$ _____

d $\dfrac{11.52}{\sqrt{0.64}}$ _____

18 **a** Find the value
of x in this diagram.
_____

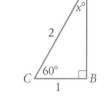

b Find the length of AB
as a surd. _____

19 Find the value of r in the diagram below.

20 Find the value of b in the diagram below.

21 Solve:

a $7 = \dfrac{x}{5}$ _____

b $\dfrac{h}{2} = 10$ _____

c $6 = \dfrac{30}{d}$ _____

d $4 = \dfrac{10}{y}$ _____

e $8 = \dfrac{18}{p}$ _____

22 Which side of this triangle is the hypotenuse?

   _____

23 If $\angle Z = a°$ in the diagram below, write an expression for the size of $\angle Y$. _____

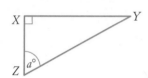

24 Round 18 min 42 s to the nearest minute.

_____

25 Find the value of x in each diagram.

a

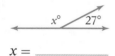

$x =$ _____

b

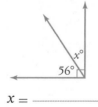

$x =$ _____

26 $\triangle STU$ is an enlargement of $\triangle PQR$.

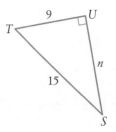

a Find the value of n. _____

b Is $\angle P$ equal to $\angle S$ or $\angle T$? _____

27 Convert 7.15 hours to hours and minutes.

_____

28 a Find the value of d in the triangle below.

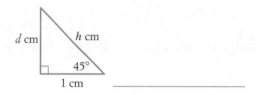

b Find the value of h, to 2 decimal places.

_____

PART C: CHALLENGE   Bonus   / 3 marks

You are given 12 coins that look identical, but one is a counterfeit (fake) and weighs *less* than the others. You are given a balance scale. Can you find the counterfeit coin after only 3 weighings?

④ TRIGONOMETRIC CALCULATIONS

FOR THIS WORKSHEET, YOU NEED TO KNOW 2 THINGS:
ONE DEGREE = 60 MINUTES, AND HOW TO USE
THE SIN, COS AND TAN KEYS ON YOUR CALCULATOR.

1 Write each angle size correct to the nearest degree.

a 64.6231° _____

b 45.5847° _____

c 18°7' _____

d 29°21' _____

e 73°19' _____

f 59°44' _____

2 Write each angle size correct to the nearest minute.

a 18°34'10" _____

b 79°6'49" _____

c 40°13'35" _____

d 36°48'22" _____

e 60°59'30" _____

f 8°14'41" _____

3 Convert to degrees and minutes, correct to the nearest minute.

a 17.5° _____

b 3.4° _____

c 46.13° _____

d 32.7° _____

e 70.21° _____

f 81.9° _____

g 73.8125° _____

h 22.4013° _____

i 49.3729° _____

j 56.4217° _____

4 Convert to decimal degrees, correct to 2 decimal places.

a 8°15' _____

b 26°12' _____

c 37°37' _____

d 13°21' _____

e 40°18' _____

f 77°4' _____

5 Calculate, correct to 3 decimal places:

a cos 64.2° _____

b tan 16.34° _____

c sin 79° _____

d tan 59°30' _____

e sin 30° _____

f 13 sin 32° _____

g 7 tan 45.7° _____

h 8 cos 27°59' _____

i 16 sin 87°8' _____

j 3 tan 8.5° _____

k $\dfrac{19}{\sin 45°}$ _____

l $\dfrac{4}{\tan 31.25°}$ _____

m $\dfrac{13.4}{\sin 51°12'}$ _____

n $\dfrac{10}{\cos 39°19'}$ _____

o $\dfrac{3}{\tan 45°}$ _____

p $\dfrac{8}{\cos 16°7'}$ _____

6 True or false? Check using a calculator.

a $\sin 60° = \dfrac{1}{2}$ _____

b $\cos 45° = \dfrac{1}{\sqrt{2}}$ _____

c $\tan 30° = \dfrac{\sqrt{2}}{2}$ _____

d $\tan 45° = 1$ _____

e $\sin 30° = \dfrac{1}{2}$ _____

f $\cos 30° = \dfrac{\sqrt{3}}{2}$ _____

g $\cos 60° = \sqrt{3}$ _____

h $\tan 60° = \sqrt{3}$ _____

i $\sin 45° = \dfrac{1}{2}$ _____

j $\sin 90° = 1$ _____

7 Find the size of angle θ, correct to the nearest degree.

a $\tan \theta = 0.467$ _____

b $\cos \theta = 0.814$ _____

c $\sin \theta = 0.204$ _____

d $\cos \theta = \dfrac{14}{18}$ _____

e $\tan \theta = \dfrac{7}{10}$ _____

f $\sin \theta = \dfrac{3}{13}$ _____

8 Find the size of angle A, correct to 2 decimal places:

a $\sin A = \dfrac{4}{5}$ _____

b $\cos A = 0.345$ _____

c $\tan A = 1.623$ _____

d $\sin A = 0.15$ _____

e $\cos A = \dfrac{5}{8}$ _____

f $\tan A = \dfrac{12}{5}$ _____

9 Find the size of angle X, correct to the nearest minute.

a $\cos X = 0.294$ _____

b $\tan X = \dfrac{9}{10}$ _____

c $\sin X = 0.73$ _____

d $\tan X = 13$ _____

e $\sin X = \dfrac{3}{5}$ _____

f $\cos X = 0.1$ _____

④ FINDING AN UNKNOWN SIDE (2)

TO SOLVE THESE PROBLEMS, YOU NEED TO KNOW WHICH SIDES
OF THE TRIANGLE ARE INVOLVED, CHOOSE THE RIGHT
TRIG RATIO, WRITE AN EQUATION, THEN SOLVE
THE EQUATION. THERE ARE MIXED ANSWERS NEXT PAGE.

WORKSHEET
WS

For each triangle, find the length of the unknown side marked, correct to 2 decimal places.

1

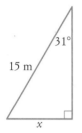

$x = $ _____

2

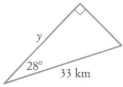

$y = $ _____

3

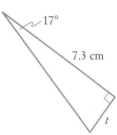

$t = $ _____

4

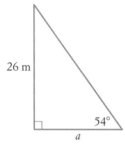

$a = $ _____

5

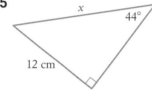

$x = $ _____

6

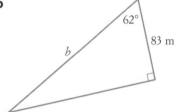

$b = $ _____

7

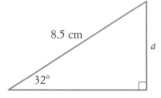

$a = $ _____

8

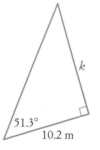

$k = $ _____

9

$x = $ _____

10

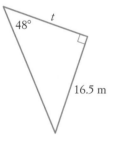

$t = $ _____

11

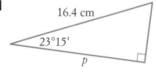

$p = $ _____

12

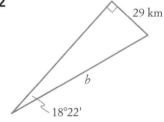

$b = $ _____

9780170454537

13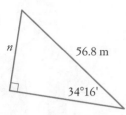

n 56.8 m 34°16'

$n =$ _____

14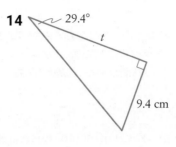

29.4° t 9.4 cm

$t =$ _____

15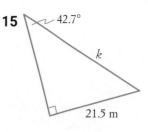

42.7° k 21.5 m

$k =$ _____

16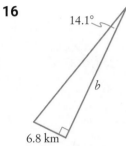

14.1° b 6.8 km

$b =$ _____

17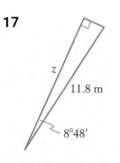

z 11.8 m 8°48'

$z =$ _____

18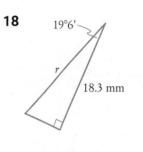

19°6' r 18.3 mm

$r =$ _____

19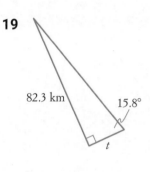

82.3 km 15.8° t

$t =$ _____

20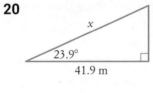

x 23.9° 41.9 m

$x =$ _____

21

31°15' k 50.2 m

$k =$ _____

WORKSHEET WS

BY THE WAY, THE ANGLES ARE MARKED WITH THE GREEK LETTERS THETA, PHI AND ALPHA. MIXED ANSWERS NEXT PAGE.

For each triangle, find the size of the angle marked, correct to the nearest degree.

1

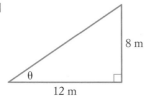

8 m

12 m

θ = _____

2

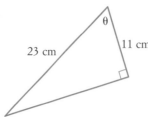

θ

23 cm

11 cm

θ = _____

3

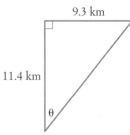

9.3 km

11.4 km

θ

θ = _____

4

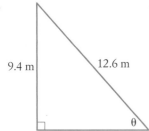

9.4 m

12.6 m

θ

θ = _____

5

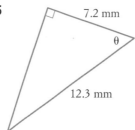

7.2 mm

θ

12.3 mm

θ = _____

6

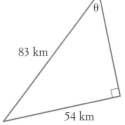

θ

83 km

54 km

θ = _____

7

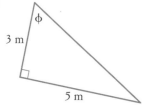

φ

3 m

5 m

φ = _____

8

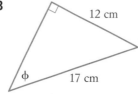

12 cm

φ

17 cm

φ = _____

9

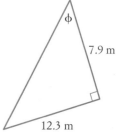

φ

7.9 m

12.3 m

φ = _____

10

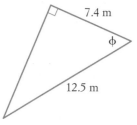

7.4 m

φ

12.5 m

φ = _____

11

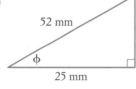

52 mm

φ

25 mm

φ = _____

12

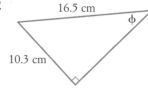

16.5 cm

φ

10.3 cm

φ = _____

WORKSHEET   WS

13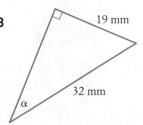

19 mm

32 mm

α

α = _____

14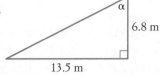

α

6.8 m

13.5 m

α = _____

15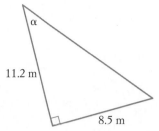

α

11.2 m

8.5 m

α = _____

16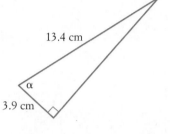

13.4 cm

α

3.9 cm

α = _____

17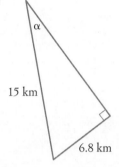

α

15 km

6.8 km

α = _____

18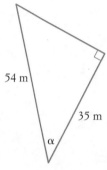

54 m

35 m

α

α = _____

19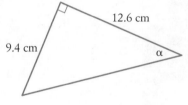

12.6 cm

9.4 cm

α

α = _____

20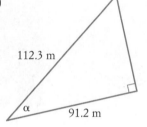

112.3 m

α

91.2 m

α = _____

21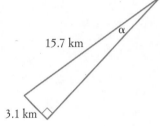

15.7 km

α

3.1 km

α = _____

ZINA HERE WITH THE WORD PUZZLE. WHEN MY DAD FOUND OUT I WAS LEARNING TRIGONOMETRY, HE SAID 'WHAT DO YOU CALL A MAN SUNBAKING AT THE BEACH? A TAN GENT.' MS LEE FOUND IT FUNNY.

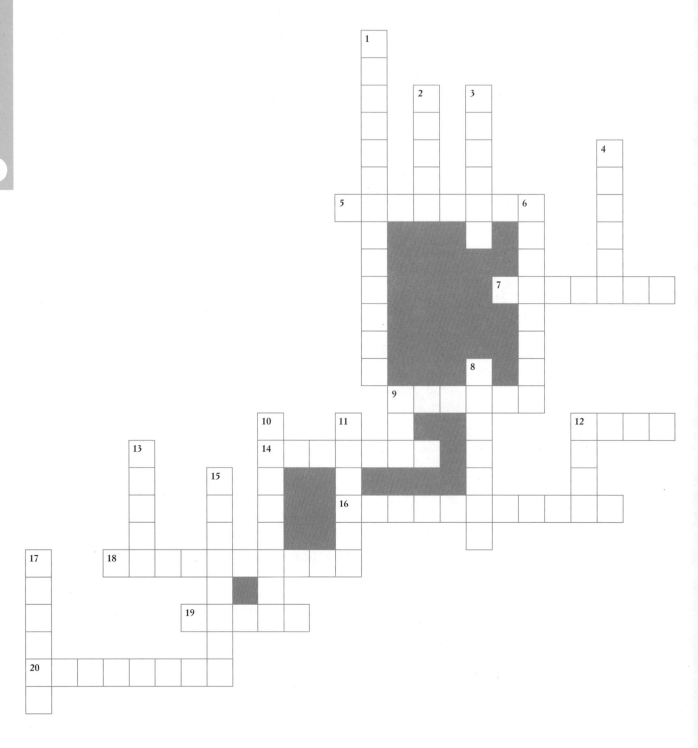

Clues across

5 The side next to an angle

7 'tan' is short for this

9 The trigonometric ratio that is $\dfrac{\text{adjacent}}{\text{hypotenuse}}$

12 The 'S' in SOHCAHTOA

14 To solve $\sin \theta = \dfrac{1}{2}$, use i _____ sine

16 A triangle with a 90° angle

18 The side opposite the 90° angle in a triangle

19 θ is a variable used to stand for the size of an _____

20 $\tan = \dfrac{\rule{1cm}{0.4pt}}{\text{adjacent}}$

Clues down

1 'P' word meaning 'crossing at 90°'

2 The Greek letter θ

3 'V' word meaning 'corner of a shape'

4 A unit for measuring angle sizes smaller than a degree

6 Three-sided polygon

8 Triangles that have the same shape but not the same size

9 The abbreviation for cosine

10 An interval inside a shape that joins 2 vertices

11 Represented by the symbol °

12 Trigonometry can be used to find the length of an unknown _____ in a right-angled triangle

13 The number of minutes in a degree

15 A word meaning 'how far'

17 sin, cos and tan are called trigonometric r_____

(4) TRIGONOMETRY 1

TRIGONOMETRY IS USED IN ENGINEERING, CONSTRUCTION AND NAVIGATION. HAVE YOU LEARNED A WAY OF REMEMBERING WHAT SIN, COS AND TAN ARE?

Name:

Due date:

Parent's signature:

Part A	/ 8 marks
Part B	/ 8 marks
Part C	/ 8 marks
Part D	/ 8 marks
Total	/ 32 marks

HW HOMEWORK

PART A: MENTAL MATHS

🖩 Calculators not allowed

1 Find the highest common factor of ac^2 and ab.

_____

2 Expand $-(y + 3)$.

_____

3 Convert 3.25% to a decimal. _____

4 Find a.

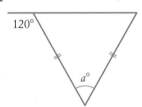

_____

5 Decrease $120 by 15%.

_____

6 Do the diagonals of a rhombus bisect each other at right angles?

_____

7 For this set of data, find:

20, 14, 4, 17, 12, 11, 18, 13

a the median

_____

b the outlier

_____

PART B: REVIEW

1 **a** Which side of this triangle is the hypotenuse?

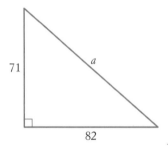

_____

b Write Pythagoras' theorem for this triangle.

_____

c Find a correct to 2 decimal places.

_____

_____

2 Simplify $\dfrac{18}{28}$. _____

3 Evaluate correct to 2 decimal places:

a $5 \times \sqrt{35}$ _____

b $\dfrac{27}{\pi}$ _____

4 Solve each equation.

a $\dfrac{x}{12} = 7.5$

_____

_____

9780170454537

b $\dfrac{7}{d} = 10$

_____

_____

PART C: *PRACTICE*

 The trigonometric ratios

1 For the marked angle, write the length of the:

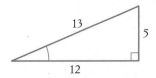

a adjacent side _____

b hypotenuse _____

c opposite side _____

2 Name the trigonometric ratio that is:

a $\dfrac{\text{adjacent}}{\text{hypotenuse}}$ _____

b $\dfrac{\text{opposite}}{\text{adjacent}}$ _____

3 Convert 30.15° to degrees and minutes.

_____

4 Write as a fraction:

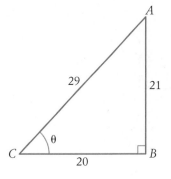

a sin θ _____

b cos θ _____

PART D: *NUMERACY AND LITERACY*

1 How many minutes in $\dfrac{1}{2}°$?

_____

2 Complete each equation with a trigonometric ratio:

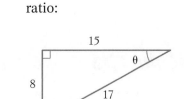

a _____ θ $= \dfrac{15}{17}$

b _____ θ $= \dfrac{8}{15}$

c _____ θ $= \dfrac{8}{17}$

3 Round each angle to the nearest degree:

a 49° 28′ _____

b 7° 40′ _____

4 Evaluate correct to 2 decimal places:

a cos 51° _____

b 20 tan 37° _____

4 TRIGONOMETRY 2

WHEN SOLVING A TRIGONOMETRY PROBLEM, YOU HAVE TO LOOK AT WHAT SIDES OF THE TRIANGLE ARE INVOLVED, WHAT TRIG RATIO TO USE, WRITE AN EQUATION AND SOLVE THE EQUATION.

Part A		/ 8 marks
Part B		/ 8 marks
Part C		/ 8 marks
Part D		/ 8 marks
Total		/ 32 marks

PART A: MENTAL MATHS

Calculators not allowed

1 Solve $\dfrac{x}{6} = 12$. _____

2 Evaluate:

a 40% of $60 _____

b $\sqrt[3]{64}$ _____

c $\dfrac{2}{3} + \dfrac{1}{2}$ _____

d $-8 \times (-3) \times 15$ _____

3 What is the probability of rolling a prime number on a die?

_____

4 Expand $(2x + 3)(2x - 3)$.

_____

_____

5 Find the perimeter of this shape.

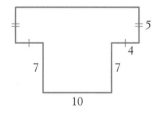

_____

PART B: REVIEW

1 In $\triangle ABC$, name the side that is:

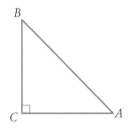

a opposite B _____

b adjacent to B _____

c adjacent to A _____

d the hypotenuse _____

2 In $\triangle LMN$, find as a fraction:

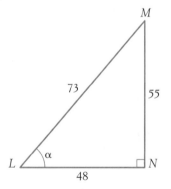

a $\cos \alpha$ _____

b $\sin \alpha$ _____

3 Evaluate correct to 2 decimal places:

a $7 \sin 18°$ _____

b $\dfrac{4}{\tan 50°11'}$ _____

HOMEWORK

HW

PART C: PRACTICE

› Finding an unknown side
› Finding an unknown angle

1 Find the value of each variable, correct to 2 decimal places.

a

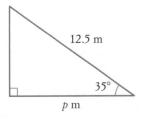
12.5 m
35°
p m

_____

_____

b

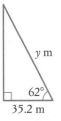
y m
62°
35.2 m

_____

_____

c

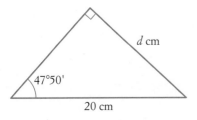
d cm
47°50'
20 cm

_____

_____

d

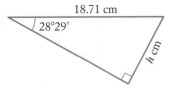
18.71 cm
28°29'
h cm

_____

_____

2 Find, correct to the nearest degree, the size of each marked angle.

a

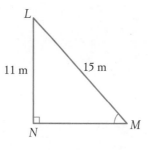
L
11 m
15 m
N
M

_____

b

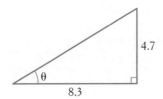
4.7
θ
8.3

_____

3 Find α in the triangle in question **2** of Part B, correct to the nearest minute.

_____

4 Find ∠*B*, correct to one decimal place.

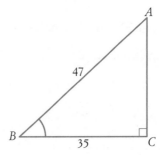
A
47
B
35
C

_____

PART D: NUMERACY AND LITERACY

1 What does sine mean?

_____

2 A car travels 4 km along a road that rises at an angle of 10°. What is the vertical height of the car, correct to the nearest metre?

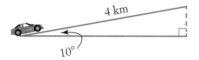

4 km

10°

_____

3 Write 36.45 in degrees and minutes.

_____

4 Write 85° 12′ in decimal degrees.

_____

5 $\triangle ABC$ is right-angled at C, $AB = 24.5$ m and $\angle B = 75°$. Find the length of side AC, correct to 2 decimal places.

_____

6 A tree casts a shadow 15 m long. If the sun's rays meet the ground at 20°, find the height of the tree correct to 2 decimal places.

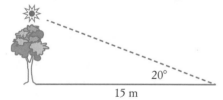

20°

15 m

_____

7 What does **adjacent** mean?

_____

8 In $\triangle TEB$, $\angle B = 90°$, $\angle T = 88°\ 45′$ and $BT = 4.3$ m.

Find BE correct to one decimal place.

_____

9780170454537

Part A	/ 8 marks
Part B	/ 8 marks
Part C	/ 8 marks
Part D	/ 8 marks
Total	/ 32 marks

TRIGONOMETRY (4) REVISION

HAVE YOU MASTERED YOUR TRIGONOMETRY YET? THERE ARE NO SHORTCUTS. YOU JUST HAVE TO LEARN THE PROCESS AND KEEP PRACTISING. GAME ON!

HW HOMEWORK

PART A: MENTAL MATHS

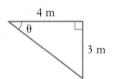 Calculators not allowed

1 Simplify $\dfrac{25ab^2c}{30bc^2}$. _____

2 For this data, find:

1	0 1 2
2	4
3	9
4	3 6

a the median _____

b the range _____

3 Evaluate:

a $41.32 + 11.50 _____

b 75% of $48 _____

4 For this triangle, find:

4 m
θ
3 m

a the perimeter _____

b the area _____

5 Expand $(2p + 1)(3p - 6)$.

_____

_____

PART B: REVIEW

1 Find θ, correct to the nearest degree.

a
28 m
36 m
θ

b
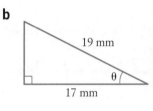
19 mm
17 mm
θ

_____

2 Find the value of each variable, correct to 2 decimal places.

a
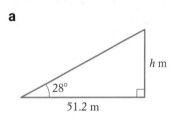
28°
51.2 m
h m

b
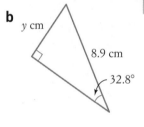
y cm
8.9 cm
32.8°

_____

_____

c
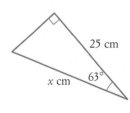
25 cm
x cm
63°

d
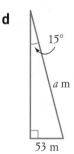
15°
a m
53 m

_____

_____

3 Find θ in question **4** of Part A, correct to the nearest minute.

_____

4 Find θ, correct to one decimal place.

48.3

θ

60.5

_____

PART C: PRACTICE

📝 › **Trigonometry revision**

1 Find *x*, correct to one decimal place.

a

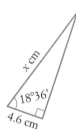

x cm

18°36′

4.6 cm

b

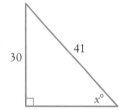

30   41

x°

_____

_____

c

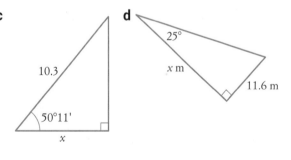

10.3

50°11′

x

d

25°

x m

11.6 m

_____

_____

2 A tree of height 15.8 m casts a shadow of length 4.1 m. What angle (to the nearest degree) do the sun's rays make with the ground?

_____

3 The angle of a missile is 32° 14′ to the ground and its line-of-sight distance is 52 km. How high is the missile, correct to one decimal place?

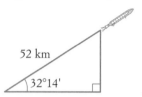

52 km

32°14′

_____

4 A surfer on the water sees an observation tower 11.8 m high at an angle of 9.5°. How far is the surfer from the tower, to 2 decimal places?

9.5°   11.8

_____

5 A tent pole 13 m high is supported by a rope 20 m long. What angle (to the nearest degree) does the rope make with the ground?

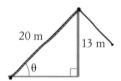

20 m   13 m

θ

_____

PART D: NUMERACY AND LITERACY

1 In △*PQR*, show that:

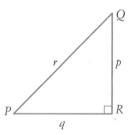

Q

r   *p*

P   *R*

q

a sin *P* = cos *Q*

_____

b tan *P* × tan *Q* = 1

_____

2 The second triangle is a reduction of the first triangle. Find, correct to the nearest whole number, the value of:

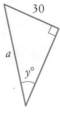

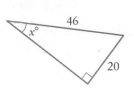

a x _____

b a _____

c y _____

3 (2 marks) Find, correct to one decimal place, all unknown sides and angles of this triangle.

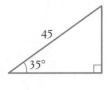

_____

_____

_____

_____

4 Use the tan ratio to prove that the equal angles in an isosceles right-angled triangle must be 45°.

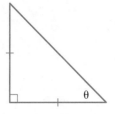

_____

_____

(5) STARTUP ASSIGNMENT 5

THIS CHAPTER IS CALLED 'INDICES', WHICH IS ANOTHER NAME FOR 'POWERS', FOR REPEATED MULTIPLICATION. IN 3^5, 3 IS CALLED THE BASE AND 5 IS CALLED THE INDEX OR POWER.

WS WORKSHEET

PART A: BASIC SKILLS   / 15 marks

1 Increase $60 by 40%. _____

2 13.5 cm = _____ mm

3 Find the value of k in this diagram.

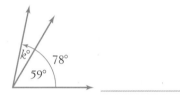

_____

4 Simplify $\dfrac{15xy}{5y}$. _____

5 If $g = 9$, evaluate $2g^2$. _____

6 Which is greater: 7.65 or 7.7? _____

7 A car travels at 70 km/h. How far will it travel in 90 minutes? _____

8 Simplify $3(2 - x) + 3x$. _____

9 Calculate $\sqrt{5 + 7}$, correct to 3 decimal places.

_____

10 Name this shape.

_____

11 Write $3 + \dfrac{7}{100} + \dfrac{1}{1000}$ as a decimal. _____

12 Calculate the mean of 6, 8, 5, 1, 0. _____

13 For this rectangle,

Area = 18 cm^2.

Width = _____

8 cm

14 Find the value of m in this diagram.

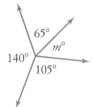

_____

15 Expand and simplify $(a + 4)(a + 5)$.

_____

_____

PART B: POWERS   / 25 marks

16 Calculate:

a $3^5 =$ _____

b $17^3 =$ _____

c $(-5)^4 =$ _____

d $(-2)^3 =$ _____

e $3.4 \times 1000 =$ _____

f $3.4 \div 1000 =$ _____

g $\sqrt{576} =$ _____

h $\sqrt{650.25} =$ _____

i $\sqrt{2^2 \times 2^4} =$ _____

j $\sqrt[3]{216} =$ _____

17 Simplify each expression in index form.

a $10^2 \times 10^3 =$ _____

b $3^7 \times 3 =$ _____

c $2^6 \div 2 =$ _____

d $(3^3)^2 =$ _____

18 Simplify:

a $b \times b$ _____

b $m \times m \times m \times m$ _____

c $n^2 \times n$ _____

d $3x^2 \times 2x^2$ _____

e $a^3 \div a^2$ _____

f $\dfrac{15m}{3m}$ _____

g $(u^2)^2$ _____

h $(2f^4)^2$ _____

19 Simplify:

a $3.75 \times 10^4 =$ _____

b $9.2 \times 10^3 =$ _____

c $8 \div 10^2 =$ _____

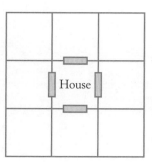

PART C: CHALLENGE   Bonus   / 3 marks

Grandad lives in a farmhouse surrounded by 8 square fields.

The house has a window in each wall and 3 fields can be seen from each window.

One day, the grandchildren visited and played in the fields. Grandad found that he could see exactly 9 children through each window. If there were 22 children playing, how many children were in each field?

(5) INDEX LAWS REVIEW

THIS WORKSHEET REVISES THE INDEX LAWS. SPEND SOME TIME ANSWERING THE QUESTIONS AND PRACTISING YOUR SKILLS.

WORKSHEET

WS

The index laws

$$a^m \times a^n = a^{m+n}$$

$$a^m \div a^n = \frac{a^m}{a^n} = a^{m-n}$$

$$(a^m)^n = a^{mn}$$

$$a^0 = 1$$

$$a^{-1} = \frac{1}{a}$$

$$a^{-2} = \frac{1}{a^2}$$

$$a^{-n} = \frac{1}{a^n}$$

1 Simplify:

a $u^4 \times u^4$ _____

b $x^2 \times x^3$ _____

c $b^5 \times b$ _____

d $t^3 \times t^4$ _____

e $6r^3 \times 4r^3$ _____

f $2m^4 \times 9m^{-1}$ _____

g $(-5a^3) \times 2a^5$ _____

h $4d^7 \times (-d^6)$ _____

i $8e^3 \times (-2e)$ _____

j $(-4d) \times (-9d^5)$ _____

2 Simplify:

a $f^7 \div f^5$ _____

b $k^6 \div k^3$ _____

c $q^{10} \div q^2$ _____

d $v^6 \div v$ _____

e $30p^6 \div 3p^5$ _____

f $24r^8 \div (-4r^3)$ _____

g $\dfrac{16d^2e^5}{2d^2e^2}$ _____

h $\dfrac{27u^6y^3}{-9u^3y^2}$ _____

i $\dfrac{-20p^7z^4}{5p^2z}$ _____

j $\dfrac{36a^5}{4a^7}$ _____

k $\dfrac{12h^2k}{15h^3}$ _____

l $\dfrac{-8c^4d^2}{12c^4d^4}$ _____

3 Simplify:

a $(r^4)^5$ _____

b $(d^3)^2$ _____

c $(u^3)^3$ _____

d $(a^2)^7$ _____

e $(t^2)^{-1}$ _____

f $(y^4)^{\frac{1}{2}}$ _____

g $(2b^3)^3$ _____

h $(3r^2)^3$ _____

i $(5u^5)^{-1}$ _____

j $(-m^2)^4$ _____

k $(-2w^6)^2$ _____

l $(4d^2)^{-2}$ _____

4 Simplify:

a z^0 _____

b 3^0 _____

c $(-10)^0$ _____

d x^0 _____

e $3b^0$ _____

f $2d^0$ _____

9780170454537

g $(3b)^0$ _____

h $(2d)^0$ _____

i $(-k)^0$ _____

j ab^0 _____

k $-7h^0$ _____

l $\left(\dfrac{1}{2}\right)^0$ _____

m $2d \times d^0$ _____

n $4m^0 \times 5m$ _____

o $(x^5)^0$ _____

p -3^0 _____

5 Write with positive indices:

a r^{-2} _____

b v^{-1} _____

c m^{-3} _____

d p^{-4} _____

e $3u^{-1}$ _____

f $10t^{-4}$ _____

g $8a^{-2}$ _____

h $5d^{-3}$ _____

i $(ab)^{-1}$ _____

j $(2k)^{-2}$ _____

k $4kr^{-1}$ _____

l $3x^{-2}y^{-1}$ _____

m $(-n)^{-1}$ _____

n $(-q)^{-2}$ _____

o $-2b^{-1}$ _____

p $-d^{-3}$ _____

6 Simplify:

a $(ab)^3$ _____

b $(f^5g^2)^4$ _____

c $(-2e^2p^3)^3$ _____

d $(-km^4n^5)^5$ _____

e $\left(\dfrac{a}{2}\right)^3$ _____

f $\left(\dfrac{b^2}{5}\right)^2$ _____

g $\left(\dfrac{r^5}{p^3}\right)^4$ _____

h $\left(\dfrac{3a^6}{bc^2}\right)^5$ _____

7 Simplify, using indices:

a $\dfrac{1}{u^2}$ _____

b $\dfrac{1}{5^3}$ _____

c $3^5 \times 3^6$ _____

d $2^3 \times 2^4 \times 2^6$ _____

e $(5^4)^2$ _____

f $\dfrac{-7}{f^2}$ _____

g $\dfrac{10}{a^4}$ _____

h $3^4 \times 4^3 \times 3^2$ _____

i $\dfrac{4^{11}}{4^9}$ _____

j $\dfrac{-1}{a}$ _____

k $\dfrac{6}{a}$ _____

l $\dfrac{5}{ab}$ _____

8 Evaluate:

a 6^{-1} _____

b 5^{-2} _____

c $\dfrac{6^{12}}{6^5}$ _____

d $\left(\dfrac{3}{5}\right)^2$ _____

e 2^0 _____

f $(3^5)^2$ _____

g $\left(-\dfrac{4}{7}\right)^3$ _____

h $\left(\dfrac{2}{3}\right)^0$ _____

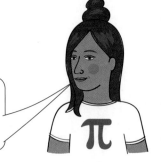

IN THIS PUZZLE, YOU HAVE TO DRAW LINES FROM THE NUMBERS
IN SCIENTIFIC NOTATION ON THE LEFT TO THEIR VALUES
ON THE RIGHT, TO DECODE THE MESSAGE NEXT PAGE.

Rule a line to connect each number in the left-hand column below to its correct decimal form in the right-hand column. If the ruled lines go through a letter and a number, the number corresponds to a number in the puzzle grid next page. Write the letters that correspond to the numbers in the grid next page to find the answer to the riddle:

How does scientific notation give you the strength of a super-hero?

2×10^3		1851
1.851×10^{-2}		0.000 000 6
6×10^3		237
6×10^{-6}		0.002
2.37×10^{-2}		0.1851
1.851×10^3		0.058
2.37×10^{-1}		6000
2×10^{-3}		18.51
5.8×10^{-3}		2370
1.851×10^{-1}		0.237
2×10^{-2}		185.1
9.1×10^8		58 000
2.37×10^2		2000
9.1×10^6		60 000
6×10^{-5}		1.851
5.8×10^{-2}		0.02
1.851×10^1		9 100 000
6×10^5		0.0237
9.1×10^7		0.000 006
5.8×10^3		91 000 000
1.851×10^0		0.018 51
2.37×10^3		23 700
5.8×10^{-4}		600 000
6×10^4		0.000 06
1.851×10^2		580 000
5.8×10^4		237 000
2.37×10^4		0.0058
6×10^{-7}		5800
5.8×10^5		910 000 000
2.37×10^5		0.000 58

Circled numbers in grid: 25, 30, 20, 28, 13, 7, 29, 22, 17, 12, 2, 18, 5, 14, 3, 19, 9, 27, 6, 1, 15, 4, 26, 24, 8, 11, 21, 16, 23, 10

Letters in grid: N, T, R, G, U, V, O, B, K, E, Y, W, I, P, S, C, N, O, W, L, Z, A, T, U, E, X, H, F, S

1	16	12	13

27	20	25

8	2	30	10

8	4	11	16

14	7	4	12	28	11	4	23	4	7

13	2	5	3	11	4	20	28

,

27	2	19	29	12	5

11	2

19	14	26

11	16	26

6	2	8	12	30	21

20	23

5	12	13

!

THIS PUZZLE WILL HELP YOU REVISE THE INDICES TOPIC IN WORDS, NOT NUMBERS. GO TO IT!

Clues across

2 Another word for 'index'

5 Anything to the power of 0 is this

6 $\dfrac{1}{a^4}$ can be written with a _____ power

7 This index always gives 1

8 Notation for writing big and small numbers

9 An educated guess

11 $(a^m)^n = a^{mn}$ is an example of an index _____

14 $3 \times 3 \times 3 \times 3$ is 3^4 written in expanded _____

15 $\dfrac{5}{2}$ is called the r _____ of $\dfrac{2}{5}$

20 The 8 in 8^3

21 0.005 019 has 4 _____ figures

22 When multiplying terms with the same base, we _____ the powers

24 The 3 in 8^3

25 439 744 rounded to 2 significant _____ is 440 000

27 From smallest to largest is _____ order

29 An 'E' word meaning power

Clues down

1 A positive or negative whole number or zero

2 The answer to a multiplication

3 The answer to a division

4 The plural of index

10 When raising a power to a power, we _____ the powers

12 7.3×10^{-5} is an example of scientific _____

13 From largest to smallest is _____ order

16 $\sqrt[3]{\ }$ means the cube _____

17 Raised to the power 3

18 Opposite of 'multiply'

19 Scientific notation uses powers of _____

23 When dividing terms with the same base, we _____ the powers

26 $r \times r \times r \times r \times r$ is written in e _____ form

28 Raised to the power 2

5 INDICES 1

INDICES IS A NEW TOPIC THAT INVOLVES NUMBER AND ALGEBRA. SPEND SOME TIME LEARNING THE INDEX LAWS AND USING THEM TO SIMPLIFY ALGEBRAIC EXPRESSIONS.

Name:

Due date:

Parent's signature:

Part A	/ 8 marks
Part B	/ 8 marks
Part C	/ 8 marks
Part D	/ 8 marks
Total	/ 32 marks

HW HOMEWORK

PART A: MENTAL MATHS

🖩 Calculators not allowed

1 Factorise $27p^2q - 15pq^2$. _____

2 Find the average of -8 and 6. _____

3 Convert $19:25$ to 12-hour time. _____

4 Evaluate:

a $\dfrac{4}{5} - \dfrac{3}{10}$

_____

b $\$40.00 - \$25.15$

_____

5 If $y = -2$, evaluate $8 - 3y$. _____

6 Decrease $60 by 15%.

_____

7 Expand and simplify $(x + 4)(3x - 4)$.

_____

_____

PART B: REVIEW

1 What is the square root of 144? _____

2 Evaluate:

a 2^5 _____

b $\sqrt{800}$ correct to 2 decimal places

_____

c $\sqrt{36 \times 64}$ _____

d $\sqrt[3]{2^9}$ _____

3 **a** Write 7^4 in words.

_____

b In 7^4, which number is the index? _____

4 What is the cube root of 125? _____

9780170454537

PART C: PRACTICE

> Multiplying and dividing terms with the same base
> Power of a power
> Powers of products and quotients

1 Simplify:

a $y^4 \times y^6$

_____

b $8x^2 \times 3x^5$

_____

c $b^{10} \div b$

_____

d $\dfrac{24r^6}{4r^2}$

_____

e $(5a^3)^2$

_____

f $\left(\dfrac{1}{2p}\right)^3$

_____

g $(6m^3n^5)^2$

_____

h $\left(\dfrac{-d^2}{ef^3}\right)^5$

_____

PART D: NUMERACY AND LITERACY

1 True or false?

a The square root of a number is always positive or zero. _____

b The cube root of a number is always positive or zero. _____

c It is not possible to find the cube root of a negative number. _____

2 What is another word for index?

_____

3 Complete: When raising a term with a power to another power, _____ the powers.

4 Evaluate:

a $\left(\dfrac{2}{7}\right)^4$ _____

b $(-1)^8$ _____

5 Write 2 terms that have a product of $-32p^7q^2$.

_____

⑤ INDICES 2

HAVE YOU LEARNED WHAT A POWER OF 0 MEANS? OR WHAT A NEGATIVE POWER IS? THIS HOMEWORK ASSIGNMENT SHOULD HELP.

Name:

Due date:

Parent's signature:

Part A	/ 8 marks
Part B	/ 8 marks
Part C	/ 8 marks
Part D	/ 8 marks
Total	/ 32 marks

PART A: MENTAL MATHS

🚫 Calculators not allowed

1 Evaluate:

a 7×30 _____

b 4^3 _____

c $33\frac{1}{3}\%$ of $24 _____

2 Write, as a decimal, the probability of selecting a heart from a standard deck of playing cards.

_____

3 Round 82.6854 to 2 decimal places.

_____

4 Mark 2 corresponding angles.

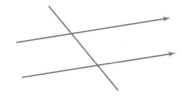

5 What percentage is $15 of $50?

_____

6 Write a simplified algebraic expression for the perimeter of a rhombus with side length d.

_____

PART B: REVIEW

1 What word describes the 5 in 5^8?

_____

2 Simplify:

a $2 \times x \times 3 \times x \times x$

_____

b $-w^6 \times (-8w^5)$

_____

c $10m^9 \div 5m^3$

_____

d $\dfrac{6c^{10}d^3}{20cd^3}$

_____

e $(3a^4)^5$

_____

f $(-3p^2qr)^7$

_____

g $\dfrac{18r^8w^5}{(2r^2w)^2}$

_____

9780170454537

PART C: PRACTICE

> Zero and negative indices
> Summary of the index laws

1 Simplify:

a $p^7 \times p^5$

_____

b y^{-2}

_____

c $8a^0$

_____

d $t^{11} \div t$

_____

e $\left(\dfrac{2}{3}\right)^{-1}$

_____

f $4u^{-3}w^{-1}$

_____

g $\dfrac{\left(3ab^2\right)^2}{6ab^5}$

_____

h $\left(-\dfrac{n}{m}\right)^0$

_____

PART D: NUMERACY AND LITERACY

1 True or false?

a Any number raised to a negative power is negative. _____

b Any number raised to a power of 0 is 1.

_____

c Any number raised to a power of -1 is its reciprocal. _____

2 Evaluate as a fraction:

a $\left(-\dfrac{1}{5}\right)^3$ _____

b $(-2)^{-4}$ _____

3 Complete:

a When dividing terms with powers,

_____ the powers.

b $(pq)^n =$ _____

4 Write 2 terms that have a quotient of $6y^4z$.

_____

5 INDICES 3

SCIENTIFIC NOTATION AND SIGNIFICANT FIGURES ARE 2 NEW THINGS COVERED IN THIS HOMEWORK ASSIGNMENT.

Name:

Due date:

Parent's signature:

Part A	/ 8 marks
Part B	/ 8 marks
Part C	/ 8 marks
Part D	/ 8 marks
Total	/ 32 marks

HW HOMEWORK

PART A: MENTAL MATHS

🔲 Calculators not allowed

1. What is $\frac{1}{3}$ as a decimal? _____

2. Evaluate $(-6)^3$ _____

3. Find f.

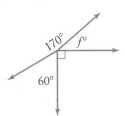

170° $f°$

60°

_____

4. Convert 0.45 to a simple fraction.

_____

5. Write an algebraic expression for the total number of legs on d dogs and p people.

_____

6. How many axes of symmetry has a parallelogram? _____

7. Are the diagonals in a parallelogram equal?

_____

8. Expand $-3(5a - 2)$ _____

PART B: REVIEW

1. Simplify:

   a $\frac{(kw)^2}{3w^3}$ _____

   b $(2p^8)^3$ _____

   c -3^{-5} _____

   d $\frac{9x^0}{24}$ _____

   e $-4m^8n \times 9mn$ _____

2. True or false:

   a $y^4 + y^3 = y^7$ _____

   b $6^{10} \div 6^5 = 1^5$ _____

3. Find the value of n if $6^n = \frac{1}{36}$.

_____

9780170454537

PART C: PRACTICE

› Significant figures
› Scientific notation

1 Express in scientific notation:

a 0.000 000 057

b 438 000 000

2 Express 26.128 correct to 3 significant figures.

3 Express 0.008 39 correct to one significant figure.

4 Simplify:

a 5.03×10^7

b 7.5×10^{-5}

5 Evaluate:

a $(9 \times 10^8) \times (4 \times 10^3)$

b $\dfrac{1.794 \times 10^7}{7.8 \times 10^4}$

PART D: NUMERACY AND LITERACY

1 Sydney had a population of 5 767 200 in 2020. Round this number to 2 significant figures.

2 Write each number in scientific notation.

a Light travels at a speed of 300 000 000 metres per second.

b The thickness of human hair is 0.000 08 m.

3 Complete:

a Scientific notation uses powers of _____ to write large and small _____.

b Zeroes at the _____ of a whole number are not significant.

c Any number raised to a power of 1 is _____.

4 a Why isn't 12×10^3 in correct scientific notation?

b Simplify 12×10^3, then rewrite the number in correct scientific notation.

HOMEWORK

HW

(5) INDICES REVISION

WE'RE AT THE END OF THE INDICES TOPIC NOW. THE INDEX LAWS AND SCIENTIFIC NOTATION SHOULD BE SINKING IN. PRACTISING THEM SHOULD HELP YOU BE A PRO AT THEM.

Name:

Due date:

Parent's signature:

Part A	/ 8 marks
Part B	/ 8 marks
Part C	/ 8 marks
Part D	/ 8 marks
Total	/ 32 marks

PART A: *MENTAL MATHS*

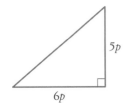 Calculators not allowed

1 Find a simplified expression for the area of this triangle.

5p

6p

_____

2 Find the value of $3x^2$ if $x = -2$. _____

3 Evaluate:

  a $6 \times 9 \times 5$ _____

  b $216 \div (-8)$ _____

  c $5 + 40 + 18 + 31 + 12$ _____

4 Simplify $8y \div 8$. _____

5 Expand $(2x + 5)(2x + 5)$.

_____

_____

6 How many hours and minutes are there between 10.35 a.m. and 4.20 p.m.?

_____

PART B: *REVIEW*

1 Round 0.078 92 to 2 significant figures.

_____

2 Write in scientific notation:

  a 0.000 514 9 _____

  b 5230 _____

  c 8.1 _____

3 Simplify 2.1×10^4. _____

4 Evaluate $(3.5 \times 10^{-3})^2$. _____

5 How many significant figures has 5470?

_____

6 Find the value of n if $8.6 \times 10^n = 0.086$.

_____

HOMEWORK

HW

9780170454537

PART C: PRACTICE

› Indices revision

1 Simplify:

a $3^0 + 4x^0$

_____

b $2y^{-3}$

_____

c $2h^2 \times 3h^6$

_____

d $10u^{18} \div 5u^9$

_____

e $\dfrac{144a^3b^2}{16b^4c^2}$

_____

2 Round 37 866 251 to 4 significant figures.

_____

3 Write 476 000 in scientific notation.

_____

4 Evaluate $\dfrac{2.1 \times 10^7 - 4.9 \times 10^5}{\left(6.36 \times 10^{-2}\right)^3}$, giving the answer in scientific notation correct to 2 significant figures.

_____

_____

PART D: NUMERACY AND LITERACY

1 Simplify $c \times c^{-1}$. _____

2 One trillion is 1 000 000 000 000.
Write this number in scientific notation.

_____

3 True or false?

a $(3^4)^3 = 3^{12}$ _____

b $5^2 \times 5^0 = 5$ _____

4 Complete:

a To _____ terms with the same base, add the powers.

b Zeroes at the _____ of a decimal are not significant.

c Any number raised to a power of _____ is 1.

5 The distance between the Earth and the Sun is 1.52×10^8 km. Write this measurement in normal decimal form.

_____

6 STARTUP ASSIGNMENT

THE GEOMETRY TOPIC COVERS THE PROPERTIES OF ANGLES, TRIANGLES AND QUADRILATERALS, SO THERE ARE SOME RULES, SYMBOLS AND TERMINOLOGY THAT YOU NEED TO KNOW.

WORKSHEET

PART A: BASIC SKILLS   / 15 marks

1 What is $\frac{3}{4}$ of 1 year? _____

2 Evaluate $(-4)^3$. _____

3 Find, the perimeter of this semi-circle, correct to 2 decimal places.

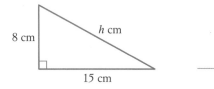
5 cm

_____

4 Find the median of these values:

8, 2, 4, 7, 3, 2, 6 _____

5 a Find the value of h in this triangle.

8 cm h cm
15 cm   _____

b Calculate the area of the triangle.

_____

6 Evaluate $\sqrt{10}$ to 3 decimal places.

_____

7 Simplify $3m \times 2m$. _____

8 Divide $440 between Dean and Jerry in the

ratio 5 : 3. _____

9 Solve the equation $2x - 7 = 1$. _____

10 Complete: 5 kg = _____ g

11 Name this shape.

12 If $x = -3$, evaluate $5x + 7$. _____

13 Decrease $52 by 8%. _____

14 Expand $4(2m - 5)$. _____

PART B: GEOMETRY   / 25 marks

15 Find the value of a in this diagram.

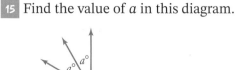

_____

16 a Mark ∠AEB in this diagram.

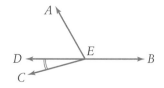

b Name the angle already marked on the

diagram. _____

17 Find the value of x.

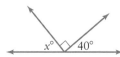
$x°$ 40°

_____

18 Find the values of a, b and c.

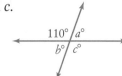
110° $a°$
$b°$ $c°$

_____

19 What is an **obtuse** angle?

_____

_____

20

s t u v
w x y z

What type of angles are:

a t and y? _____

b w and y? _____

21 How many axes of symmetry has a rectangle?

_____

22 Find the value of *w* in this triangle.

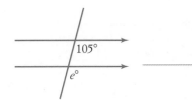

_____

23 Find the value of *e* in this diagram.

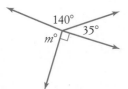

_____

24 What is the size of each angle in an

equilateral triangle? _____

25 Find the value of *m* in this diagram.

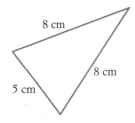

_____

26 Mark the 2 equal angles in this triangle.

8 cm

8 cm

5 cm

27 Use a protractor to draw a 135° angle.

28 Draw a triangular pyramid.

29 What is the **supplement** of 50°?

_____

30 Does a trapezium have rotational symmetry?

_____

31 Find the value of *t* in this diagram.

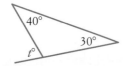
40°

30°

t°

_____

32 Draw a pair of **vertically opposite** angles.

33 Find the value of *u* in this diagram.

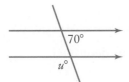
70°

u°

_____

34 Sketch a rhombus.

PART C: CHALLENGE   Bonus   / 3 marks

Use 10 coins or counters to make this triangular shape. Then, by moving only 3 coins or counters, turn the triangle upside down.

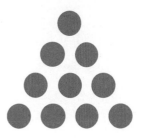

 **6 NAMING QUADRILATERALS**

THERE ARE 6 SPECIAL QUADRILATERALS, EACH WITH ITS OWN PROPERTIES OF SIDES, ANGLES AND DIAGONALS. WHICH ONES DON'T YOU KNOW?

A **trapezium** is a quadrilateral with at least one pair of opposite sides parallel.

A **kite** is a convex quadrilateral with 2 pairs of adjacent sides equal in length.

A **parallelogram** is a quadrilateral with both pairs of opposite sides parallel.

A **rhombus** is a parallelogram with 2 adjacent sides equal in length.

A **rectangle** is a parallelogram with one angle a right angle.

A **square** is a rectangle with 2 adjacent sides equal in length.

1 Use the definitions above to help you sketch each of the 6 quadrilaterals described.

In Questions **2** to **5**, state whether each statement is TRUE or FALSE.

2 a A square is a trapezium. _____

b A rectangle is a trapezium. _____

c A kite is a trapezium. _____

d A parallelogram is a trapezium. _____

e A rhombus is a trapezium. _____

3 a A square is a rhombus. _____

b A trapezium is a rhombus. _____

c A rectangle is a rhombus. _____

d A parallelogram is a rhombus. _____

e A kite is a rhombus. _____

4 a A kite is a rectangle. _____

b A rhombus is a rectangle. _____

c A square is a rectangle. _____

d A parallelogram is a rectangle. _____

e A trapezium is a rectangle. _____

5 a A trapezium is a kite. _____

b A rhombus is a kite. _____

c A parallelogram is a kite. _____

d A square is a kite. _____

e A rectangle is a kite. _____

9780170454537

6 List all quadrilaterals that have:

a opposite sides equal _____

_____

b opposite sides parallel _____

_____

7 Which quadrilaterals have:

a equal diagonals? _____

_____

b adjacent sides equal? _____

_____

c all angles equal? _____

_____

d diagonals that cross at right angles? _____

_____

e diagonals that bisect each other? _____

_____

f all sides equal? _____

_____

g diagonals that bisect each other at right angles?

_____

h diagonals that bisect the angles of the

quadrilateral? _____

i 2 axes of symmetry? _____

_____

j rotational symmetry? _____

_____

8 Name the most general quadrilateral that has:

a all sides equal _____

b diagonals equal _____

c 2 pairs of opposite sides parallel and equal

_____

d diagonals that bisect at 90°

_____

e diagonals that cross at 90° and one is bisected

_____

f all angles 90° _____

g opposite sides equal _____

h diagonals that bisect each other

_____

i one pair of opposite sides parallel

_____

j diagonals that are equal and bisect each other.

_____

6 FIND THE UNKNOWN ANGLE

- Isosceles and equilateral triangles
- Exterior angle of a triangle
- Quadrilaterals
- Regular polygons
- Angle sums

Find the value of each variable.

1

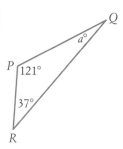

$a =$ _____

2

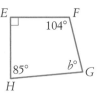

$b =$ _____

3

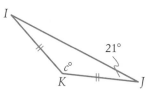

$c =$ _____

4

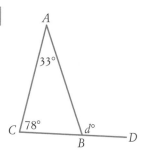

$d =$ _____

5

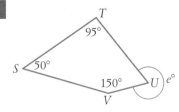

$e =$ _____

6

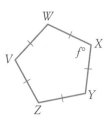

$f =$ _____

7

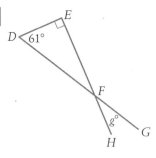

$g =$ _____

8

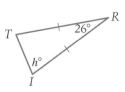

$h =$ _____

9

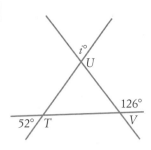

$i =$ _____

9780170454537

10

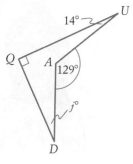

$j =$ _____

11

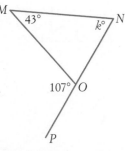

$k =$ _____

12

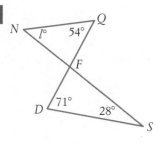

$l =$ _____

13

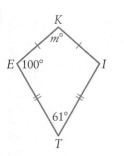

$m =$ _____

14

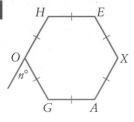

$n =$ _____

9780170454537

15

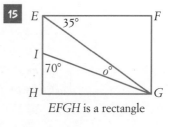

EFGH is a rectangle

$o =$ _____

16

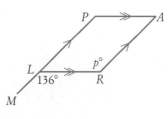

$p =$ _____

17

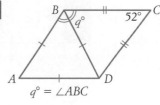

$q° = \angle ABC$

$q =$ _____

18

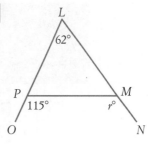

$r =$ _____

19

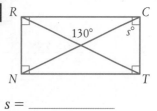

$s =$ _____

20

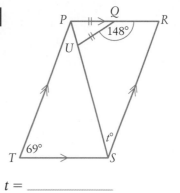

$t =$ _____

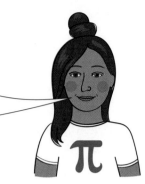

MS LEE WAS RIGHT WHEN SHE SAID THERE WAS A LOT OF TERMINOLOGY IN THIS TOPIC. SEE HOW WELL YOU KNOW YOUR GEOMETRY WORDS.

Clues across

4 Two intersecting lines create v_____ opposite angles

7 Between parallel lines crossed by a transversal, the 2 angles that are supplementary

10 Opposite _____ of a parallelogram are equal

12 A c_____ polygon has vertices that all point outwards

14 'Free' angles that add to 90°

18 A r_____ polygon has equal sides and angles

19 A triangle with no equal sides

21 A quadrilateral with one pair of parallel sides

23 A regular quadrilateral

25 A triangle with all angles equal

28 The diagonals of a kite are p_____

Clues down

1 Triangle with 2 'equal legs' (Greek)

2 Quadrilateral with opposite sides parallel

3 Another name for a line of symmetry

5 To cut in half

6 This shape is a musical instrument

8 Vertically opposite angles are _____

9 This shape you can fly

11 The number of degrees in each angle of an equilateral triangle

13 The general name for a 4-sided shape

15 How many lines of symmetry has a kite?

16 Lines that never cross

17 How many lines of symmetry has a parallelogram?

20 A triangle's _____ angle equals the sum of its 2 opposite interior angles

22 An interval joining opposite corners of a shape

24 A parallelogram with right angles

26 A kite has 2 pairs of equal _____ sides

27 A 10-sided shape

29 Any shape with straight sides

⑥ GEOMETRY 1

THIS HOMEWORK ASSIGNMENT REVISES THE GEOMETRY RULES, SUCH AS ANGLES ON A STRAIGHT LINE, CORRESPONDING ANGLES ON PARALLEL LINES AND THE ANGLE SUM OF A TRIANGLE. HOW WELL DO YOU KNOW THESE RULES?

Name:

Due date:

Parent's signature:

Part A	/ 8 marks
Part B	/ 8 marks
Part C	/ 8 marks
Part D	/ 8 marks
Total	/ 32 marks

PART A: *MENTAL MATHS*

🚫 Calculators not allowed

1 For the data in this dot plot, find:

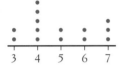

   3   4   5   6   7

a the mode _____

b the median _____

2 Simplify 35 : 14. _____

3 Evaluate $7.40 + $12.60. _____

4 Find the volume of this cube.

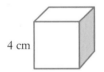

4 cm

_____

5 Increase $160 by 5%. _____

6 Simplify $\dfrac{20k^6}{4k}$. _____

7 Expand and simplify $2(4x - 4) + 6(x + 2)$.

_____

_____

PART B: *REVIEW*

1 Name each type of angle.

a   **b**

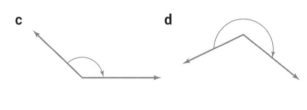

_____   _____

c   **d**

_____   _____

2 What is:

a the angle sum of a triangle? _____

b the sum of angles on a straight line?

_____

c the sum of cointerior angles on parallel lines?

_____

d the size of each angle in an equilateral

triangle? _____

C
S
F

9780170454537

PART C: PRACTICE

1 Find the value of each variable.

a

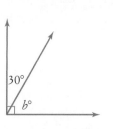

30°

b°

_____

_____

b

d° 75°

_____

_____

c

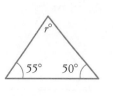

r°

55° 50°

_____

_____

d

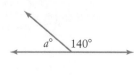

a° 140°

_____

_____

e

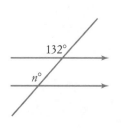

132°

n°

_____

_____

f

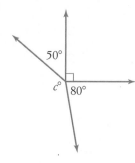

50°

c° 80°

_____

_____

g

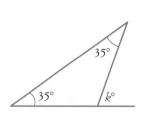

35°

35° *k*°

_____

_____

h

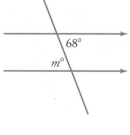

68°

m°

_____

_____

PART D: NUMERACY AND LITERACY

1 Complete:

a The exterior angle of a triangle is equal to the sum of

_____

_____

b Vertically opposite angles are _____

c Complementary angles add up to _____

d Angles at a point add up to _____

2 Mark a pair of alternate angles.

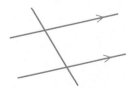

3 How many axes of symmetry has a scalene triangle? _____

4 How many acute angles are in an obtuse-angled triangle? _____

5 Why isn't *AB* parallel to *CD*?

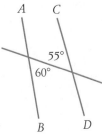

A *C*

55°
60°

B *D*

_____

_____

6 GEOMETRY 2

MATHS ISN'T JUST ABOUT NUMBERS. YOU HAVE TO DO SOME THINKING AND PROBLEM-SOLVING IN GEOMETRY, AND USE THE CORRECT WORDS TO EXPLAIN YOUR ANSWERS.

Name:

Due date:

Parent's signature:

Part A	/ 8 marks
Part B	/ 8 marks
Part C	/ 8 marks
Part D	/ 8 marks
Total	/ 32 marks

HOMEWORK

PART A: MENTAL MATHS

🚫 Calculators not allowed

1 Write Pythagoras' theorem for this triangle.

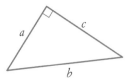

_____

2 Evaluate:

a $17 \times 5 \times 4$ _____

b $-9 + 5 + (-6)$ _____

3 Complete: $\dfrac{2}{5} = \dfrac{\square}{25}$. _____

4 Find the area of this shape.

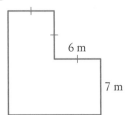

6 m

7 m

_____

5 Convert 28% to a simple fraction. _____

6 Simplify $\left(\dfrac{3x^5}{4}\right)^2$. _____

7 Which trigonometric ratio is $\dfrac{\text{opposite}}{\text{hypotenuse}}$?

_____

PART B: REVIEW

1 Find the value of each variable.

a

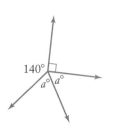

140°

$a°$ $a°$

_____
_____

b

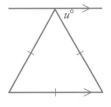

$u°$

_____
_____

c

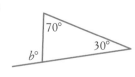

70°

$b°$ 30°

_____
_____

d

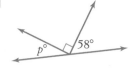

$p°$ 58°

_____
_____

e

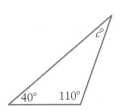

$c°$

40° 110°

_____
_____

f

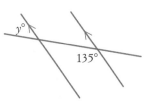

$y°$

135°

_____
_____

g

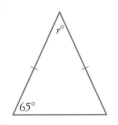

$r°$

65°

_____
_____

h

152° $x°$

_____
_____

86   Nelson Maths Workbook 3
9780170454537

PART C: PRACTICE

1 Name the quadrilateral that matches each definition.

a A special type of parallelogram with all angles equal

_____

b All sides are equal _____

c All sides are equal and all angles are equal

_____

d 2 pairs of equal adjacent sides

_____

2 Find the value of each variable.

a   **b**

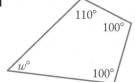   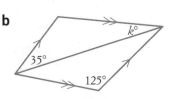

_____   _____

_____   _____

c   **d**

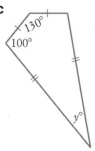   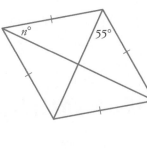

_____   _____

_____   _____

PART D: NUMERACY AND LITERACY

1 a What is a parallelogram?

_____

_____

b Which angles in a parallelogram are equal?

_____

c Write one property of the diagonals of a parallelogram.

_____

2 Write algebraically for this triangle the rule that the exterior angle is equal to the sum of the 2 interior opposite angles.

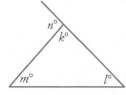

_____

3 a Name this special quadrilateral.

_____

b Find *l*, giving a reason.

_____

_____

4 a Draw a trapezium.

b How many axes of symmetry has a trapezium?

_____

STARTUP ASSIGNMENT 7

THIS ASSIGNMENT REVISES PREVIOUS WORK ON EQUATIONS TO PREPARE YOU FOR THE NEW TOPIC. REMEMBER THAT YOU CAN CHECK YOUR SOLUTIONS BY SUBSTITUTING BACK INTO THE EQUATION.

PART A: BASIC SKILLS   / 15 marks

1 Decrease $35 by 35%. _____

2 Expand and simplify $7(2p + 1) - 3(1 - 2p)$.

_____

_____

3 Simplify $2g^3 \times 5g$. _____

4 Evaluate $(6.01 \times 10^5) \times (2 \times 10^{-7})$.

_____

5 a Find the value of h in the triangle below.

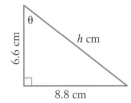

_____

b Find the value of θ in the above triangle, correct to the nearest degree. _____

6 Evaluate $\dfrac{3}{5} - \dfrac{3}{8}$. _____

7 Evaluate $\dfrac{(5.6 - 8.1)^2}{0.38 + 0.87}$. _____

8 Write $\dfrac{5}{6}$ as a decimal. _____

9 Write one property of the diagonals of a rhombus.

_____

_____

10 Find the value of y in this quadrilateral.

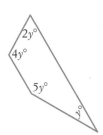

_____

11 Louise earns $28.42 per hour. Calculate her pay if she works for $7\dfrac{1}{2}$ hours. _____

12 Expand $(3x - 4)(x + 5)$.

_____

_____

13 A rectangle is twice as long as it is wide and has a perimeter of 72 cm. What is its area?

_____

_____

14 Calculate, correct to 3 significant figures, the volume of this cylinder.

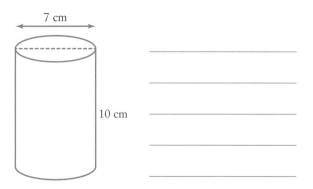

_____

_____

_____

_____

_____

PART B: EQUATIONS   / 25 marks

15 Test whether $x = 3$ is the solution to:

a $8x - 7 = 17$

_____

b $4(3 - 2x) = x - 15$

_____

_____

9780170454537

16 Test if $m = -1$ is the solution to:

a $\dfrac{m+3}{2} = 2$

_____

b $2m + 10 = 3m - 6$

_____

17 Write any equation whose solution is:

a $x = 7$

_____

b $p = 4$

_____

c $u = -3$

_____

d $r = \dfrac{1}{2}$

_____

18 Find the lowest common multiple (LCM) of:

a 8 and 10 _____

b 5 and 4 _____

c 4 and 8 _____

19 Simplify:

a $7(2k - 1)$ _____

b $-3(3k - 5)$ _____

c $\dfrac{x}{4} \times 8$ _____

d $\dfrac{3d}{2} \times 10$ _____

e $6 \times \dfrac{4 - 3t}{3}$ _____

20 Write an algebraic expression for:

a h decreased by 4 _____

b 1 more than double x _____

c the difference between p and 2 _____

21 Solve each equation.

a $x + 8 = 4$

_____

b $3p - 14 = 25$

_____

_____

c $\dfrac{k}{3} = 12$

_____

d $2m - 1 = 8$

_____

_____

e $12 = 4(a - 1)$

_____

_____

f $2y + 1 = y + 3$

_____

_____

PART C: CHALLENGE   Bonus   / 3 marks

Stan bought 7 packets of lollies while Oliver bought 4 packets. Each packet had the same number of lollies. After Stan had eaten 58 of his lollies and Oliver had eaten 19 of his, both had the same number of lollies remaining. How many lollies were in each packet to begin with?

7 WORD PROBLEMS WITH EQUATIONS

> CONVERTING A WORDED PROBLEM INTO A MATHEMATICAL EQUATION IS AN IMPORTANT SKILL. YOU HAVE TO KNOW WHAT WORDS LIKE SUM, TRIPLE, AVERAGE AND CONSECUTIVE MEAN.

WORKSHEET

1 When 4 times a number is subtracted from 300, the answer is 176. What is the number?

_____

2 7 more than double a number is equal to 5 less than triple that number. What is the number?

_____

3 Dominic was 8 years old when his sister was born. Now the sum of their ages is 56. How old is Dominic now?

_____

4 The perimeter of this triangle is 30 cm. Find the length of each side.

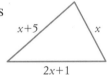

_____

5 The sum of 2 consecutive whole numbers is 99. What are the numbers? _____

6 Find the value of x in this diagram.

_____

7 Harry and Nicky share $120 but Nicky takes $58 more than Harry. How much does each person have?

_____

8 When Katy was 12, her father was 3 times her age. Now he is twice her age. How old is Katy now?

_____

9 A rectangle is 4 cm longer than it is wide. Its perimeter is 64 cm. Find its dimensions.

_____

10 The sum of 3 consecutive numbers is 99. What are the numbers?

_____

11 Mark has 28 10c and 20c coins. How many of each coin does he have if their total value is $4.30?

_____

12 The angle sum of an n-sided polygon is given by the formula $A = 180(n - 2)°$. What type of polygon has an angle sum of 1080°?

_____

13 In a class of 26 students, there are 4 more boys than girls. How many boys are there?

_____

14 This triangle has an area of 54 cm². Find the value of x.

_____

9780170454537

15 Co-interior angles in a parallelogram are $(2a + 36)°$ and $(3a − 11)°$. How big is the obtuse angle?

_____

16 A group of 30 tourists paid a total of $686 in bus fares. If the adult fare was $28 and children paid half fare, how many adults were in the group?

_____

17 The sum of 3 consecutive even numbers is 96. What are the numbers?

_____

18 Jacqueline was 24 when her son was born. Now she is 3 times her son's age. How old is Jacqueline now?

_____

19 This trapezium has an area of 22 cm². Find the value of x.

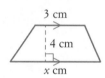

_____

20 Mr Badger bought 12 pens every month until their price rose by 15 cents each. Now he can only afford to buy 10 pens at the same total cost as before. What was the original price of each pen?

_____

21 Find the size of the smallest angle in this triangle.

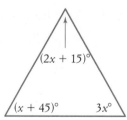

_____

22 This rectangular prism has a surface area of 184 cm². Find the value of h.

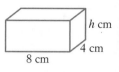

_____

23 One day, $\frac{3}{5}$ of the office staff came to work. If 3 more people had been away, this fraction would have decreased to $\frac{3}{8}$. How many people are on the full staff?

_____

24 There are 41 pigs and chickens on a farm. If there are 100 legs counted altogether, how many of each animal are there?

_____

25 Mrs Grant was 20 when her eldest child, Tess, was born. Carly was born 2 years later and Troy another 4 years later. Now the average of their 4 ages is 39. How old are Mrs Grant and her 3 children?

_____

7 WORKING WITH FORMULAS

FORMULAS ARE RULES WRITTEN USING LETTERS OF THE ALPHABET CALLED VARIABLES. THEY ARE AN IMPORTANT PART OF ALGEBRA.

1 The braking distance, d m, of a bicycle travelling at speed V m/s is $d = \dfrac{V(V+1)}{2}$.

Find the braking distance at a speed of 8.5 m/s.

_____

2 For the formula $A = LB$, find the value of:

a A when L is 12.5 and $B = 9$

_____

b L when $A = 91$ and $B = 6.5$.

_____

3 Given the formula $E = mc^2$, find E if $m = 4.7 \times 10^{-20}$ and $c = 3 \times 10^8$.

_____

4 From a height of h m, a person can see a distance of d km to the horizon, where:

$$d = 8\sqrt{\dfrac{h}{5}}$$

What distance (to the nearest kilometre) can be seen from the top of Sydney Tower, 304 m high?

_____

5 The formula for converting feet to metres is:

$$M = 0.3048F$$

Convert, correct to 2 decimal places:

a 6 ft to metres

_____

b 10 m to feet.

_____

6 The number of matchsticks, m, used to make this pattern of triangles is $m = 2t + 1$, where t is the number of triangles.

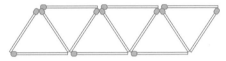

How many:

a matches are needed to make 65 triangles?

_____

b triangles can be made with 55 matches?

_____

7 a Find a formula for calculating the number of matchsticks, m, needed to make a pattern of s squares.

_____

b Use your formula to find the number of matchsticks needed to make 15 squares.

_____

8 The area of a trapezium is $A = \dfrac{1}{2}(a + b)h$.

Find the height of a trapezium that has an area of 52.5 cm² and parallel sides of lengths 5 cm and 9 cm.

_____

9 If $S = \dfrac{n}{2}(a + l)$, find n when $S = 14\,928$, $a = 25$ and $l = 286$.

_____

10 The volume of a cylinder is $V = \pi r^2 h$. Find, correct to 3 significant figures:

a the volume of a cylinder with radius 6 m and height 10.5 m

_____

b the height of a cylinder with volume 12 211.68 cm³ and radius 8 cm.

_____

11 The formula for converting kilometres to miles is $M = \dfrac{5k}{8}$.

Use the formula to convert:

a 25 km to miles _____

b 8 miles to kilometres. _____

12 Find the radius of a circle that has a circumference of 43.98 cm.

Give your answer correct to 2 decimal places.

_____

13 Find the height of a triangle with an area of 84 m² if its base is 14 m long.

_____

14 The mean of 2 numbers, x and y, is:
$$M = \frac{x+y}{2}.$$
The mean of 78 and another number is 63.

What is the other number?

_____

15 The formula for converting Fahrenheit (°F) temperatures to Celsius (°C) is:
$$C = \frac{5}{9}(F - 32)$$

a Boiling point is 212 °F. What is this temperature in Celsius? _____

b Convert a temperature of 200 °C to °F.

_____

16 The time, T, it takes a swing to swing back and forth is $T = 2\pi\sqrt{\dfrac{l}{g}}$, where l is the length of the swing and g is the gravitational acceleration. Find T, to 2 decimal places, if $l = 2.6$ and $g = 9.8$.

_____

17 The angle sum (in degrees) of a polygon with n sides is $A = 180(n - 2)$. Find the number of sides of the polygon that has an angle sum of 1440°.

_____

18 The cost of hiring a chauffeured limousine is $185 plus $3.70 for each kilometre travelled.

a Calculate the cost of hiring a limousine to travel 22 km.

_____

b Construct a formula that calculates the cost, C, of hiring a limousine to travel d km.

_____

⑦ EQUATIONS 1

Name:

Due date:

Parent's signature:

Part A	/ 8 marks
Part B	/ 8 marks
Part C	/ 8 marks
Part D	/ 8 marks
Total	/ 32 marks

ALGEBRA CAN BE HARD IF YOU DON'T KNOW THE BASICS. LEARN THE BASICS SO THAT YOU DON'T GET LOST WHEN THE EQUATIONS GET HARDER. STAY ON TOP OF THINGS.

PART A: MENTAL MATHS

🚫 Calculators not allowed

1 Evaluate $33.2 - 3.9$.

_____

2 Round $923.674 to the nearest dollar.

_____

3 Convert 9.05% to a decimal _____

4 Name this shape and state how many axes of symmetry it has.

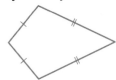

_____

5 Expand and simplify $2(x + y) - 3(x - y)$.

_____

_____

6 Find the area of this parallelogram.

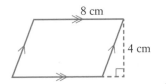

_____

7 Simplify $\dfrac{\left(t^3\right)^5}{t^4}$. _____

8 Simplify 2×10^{-3}. _____

PART B: REVIEW

1 Solve:

a $n - 8 = 32$ _____

b $6y = 30$ _____

c $\dfrac{r}{3} = 9$ _____

2 Expand:

a $4(x + 3)$ _____

b $8(d - 7)$ _____

c $-5(a - 2)$ _____

3 If $k = -6$, evaluate:

a $5k - 4$ _____

b $3(k + 10)$ _____

PART C: PRACTICE

📝 › Two-step equations
› Equations with variables on both sides

1 Solve:

a $2u + 9 = 45$

_____

_____

b $-3z + 6 = 33$

_____

_____

9780170454537

c $45 - 5h = 20$

_____

_____

d $\dfrac{2b}{5} = 18$

_____

_____

e $2a + 4 = a + 19$

_____

_____

f $5p - 10 = 2p - 25$

_____

_____

g $7 + y = 5y + 39$

_____

_____

h $15 - 3x = 1 - x$

_____

_____

PART D: NUMERACY AND LITERACY

1 **a** Write a simplified algebraic expression for the perimeter of this triangle.

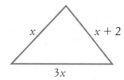

_____

b Find x if the perimeter of this triangle is 57 cm.

_____

_____

2 For $6x - 2 = 3x + 7$, explain:

a your first step in solving this equation

_____

_____

b your second step

_____

_____

c your third step

_____

_____

3 Write a two-step equation with solution $x = 5$.

_____

_____

4 Circle where the error has been made in solving this equation:

$2y - 5 = 5 - 3y$

$5y - 5 = 5$

$5y = 0$

$y = 0$

5 Describe how to check the solution to an equation.

_____

_____

⑦ EQUATIONS 2

TO CONVERT A WORD PROBLEM INTO AN EQUATION, CHOOSE A VARIABLE TO STAND FOR WHAT NEEDS TO BE FOUND, THEN USE THE CLUES TO WRITE AN EQUATION INVOLVING THE VARIABLE. THEN SOLVE THE EQUATION TO SOLVE THE PROBLEM.

Part A	/ 8 marks
Part B	/ 8 marks
Part C	/ 8 marks
Part D	/ 8 marks
Total	/ 32 marks

PART A: MENTAL MATHS

🚫 Calculators not allowed

1 Write $\dfrac{874}{1000}$ as:

   a a decimal _____

   b a percentage _____

2 Find x.

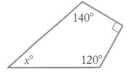

_____

3 Evaluate 6^3. _____

4 Arrange in ascending order:

   89.901, 89.9, 89.92, 89.918.

   _____

5 Complete: 1 m^2 = _____ cm^2.

6 Write 495 000 in scientific notation.

   _____

7 If a rectangle has length l and width w, find a simple algebraic expression for its perimeter.

   _____

PART B: REVIEW

1 Solve:

   a $5a - 6 = 29$

   _____

   _____

b $-3x + 7 = 9$

_____

_____

c $7 - 8h = -12$

_____

_____

d $\dfrac{r + 9}{4} = -1$

_____

_____

e $6a - 9 = 2a + 15$

_____

_____

f $5p - 28 = 3p - 12$

_____

_____

g $6 - 4t = 2 - 2t$

_____

_____

h $3u - 13 = 5u$

_____

_____

9780170454537

PART C: PRACTICE

📝 › Equations with brackets
   › Equations with algebraic fractions
   › Equation problems

1 Solve:

a $2(w + 2) = 12$

_____

_____

b $-3(d - 5) = -1$

_____

_____

c $7(2m - 1) = 3m + 16$

_____

_____

d $6(x + 4) = 2(2x + 17)$

_____

_____

e $\dfrac{a + 5}{4} = \dfrac{a + 2}{3}$

_____

_____

f $\dfrac{y}{3} + \dfrac{y}{2} = 30$

_____

_____

2 A rectangle is twice as long as it is wide.
If its perimeter is 42 m, use an equation to
find its length and width.

_____

_____

3 The sum of 2 consecutive even numbers is 70.
Use an equation to find the 2 numbers.

_____

_____

PART D: NUMERACY AND LITERACY

1 (3 marks) Find the size of each angle in this
triangle.

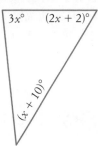

$3x°$ $(2x + 2)°$

$(x + 10)°$

_____

_____

_____

_____

_____

2 Given the formula $v = \dfrac{ct}{2}$, find v when $c = 10$
and $t = 2.5$.

_____

3 What word means the answer to an equation
or the value that makes the equation true?

_____

4 When 4 is added to 8 times a number, the
answer is 68. Find the number.

_____

_____

5 (2 marks) Kate is 3 years older than Mark.
Sarah, Kate's mother, is 4 times Kate's age.
The sum of the 3 ages is 57. How old is Kate?

_____

_____

_____

_____

C
S
F

⑦ EQUATIONS REVISION

Name:

Due date:

Parent's signature:

WE'RE UP TO THE LAST HOMEWORK ASSIGNMENT ON EQUATIONS. IF YOU CAN DO EVERY QUESTION HERE, THEN YOU HAVE ACED THE TOPIC. GOOD LUCK!

Part A	/ 8 marks
Part B	/ 8 marks
Part C	/ 8 marks
Part D	/ 8 marks
Total	/ 32 marks

PART A: MENTAL MATHS

🖩 Calculators not allowed

1 Factorise $15y - 5xy$. _____

2 Evaluate $\dfrac{7}{8} - \dfrac{2}{3}$. _____

3 Find the volume of this cube.

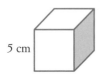

5 cm

_____

4 Write 8.1376, correct to 4 significant figures.

_____

5 Find x.

240°

$x°$

_____

6 Decrease $250 by 30%. _____

7 Expand $(3x - 5)(x - 4)$.

_____

_____

8 Divide $1800 between Rose and Khalid in the ratio 4 : 5.

_____

_____

PART B: REVIEW

1 Solve:

   a $8 - 5y = -17$

_____

_____

   b $\dfrac{2x + 2}{3} = 6$

_____

_____

   c $3n + 8 = 17$

_____

   d $\dfrac{p}{3} - \dfrac{p}{7} = 4$

_____

_____

   e $3w + 3 = 8 - 2w$

_____

_____

   f $7b + 1 = -b - 3$

_____

_____

   g $5(3 + k) = -40$

_____

_____

   h $-4(s - 2) = 2s$

_____

_____

9780170454537

PART C: PRACTICE

› Equations revision

1 Solve:

a $\dfrac{2-2x}{4}=10$

b $2(f+1)=3(2f-1)$

2 The perimeter of this rectangle is 82 m. Find:

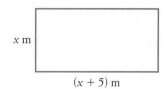

x m

$(x+5)$ m

a its length

b its area

3 (2 marks) The sum of 3 consecutive numbers is 81. Use an equation to find the 3 numbers.

4 For the formula $F=\dfrac{9C}{5}+32$, find:

a F when $C=-5$

b C when $F=86$

PART D: NUMERACY AND LITERACY

1 For $8(x-4)=32$:

a what name is given to the x?

b describe the first step for solving the equation

c describe the second step

2 Solve $\dfrac{p-2}{4}=\dfrac{2p+5}{3}$.

3 The formula for the perimeter of a rectangle of length l and width w is $P=2l+2w$.
Find w if $P=80$ and $l=33$.

4 If twice a number is subtracted from 45, the answer is 27. What is the number?

5 (2 marks) At a fun run, 509 people paid a total of $1927 to participate. Each adult paid $5, while each child paid $3. How many children participated in the fun run?

7 EQUATIONS CROSSWORD

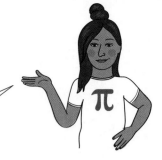

HI, IT'S ME, ZINA. THIS CROSSWORD IS UNUSUAL
IN THAT THE ANSWERS ARE GIVEN BELOW.
YOU JUST HAVE TO WORK OUT WHERE
THEY FIT IN THE CROSSWORD.

The answers to this crossword are listed below in alphabetical order. Arrange the words in the correct places on the puzzle.

ALGEBRA	CHECK	CONSECUTIVE	CUBIC
EQUATION	EQUAL	FRACTION	INVERSE
LCM	LHS	LINEAR	MULTIPLE
NUMBER	OPERATION	PRONUMERAL	QUADRATIC
RHS	ROOT	SOLVE	SQUARE
SUBJECT	SUBSTITUTE	SURD	TEST
UNDOING	UNKNOWN	VARIABLE	

HI, IT'S MS LEE. EARNING MONEY IS A TOPIC THAT USES A LOT OF OUR NUMBER SKILLS WITH PERCENTAGES, DECIMALS, RATES AND TIME. YOU WILL FIND THIS USEFUL WHEN YOU START A PART-TIME JOB.

WS

WORKSHEET

PART A: BASIC SKILLS   / 15 marks

1 Expand $a(a - 4)$. _____

2 Find the median of 7, 15, 18 and 20. _____

3 Find the area of this triangle.

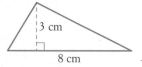

3 cm

8 cm _____

4 Simplify $x + 3y + 3x - 2y$. _____

5 Find $3 + (-8)$. _____

6 Write the value of π, correct to 4 significant figures. _____

7 Write Pythagoras' theorem for this triangle.

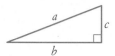

a

c

b _____

8 If $p = 7$, evaluate $12 - 2p$. _____

9 Find the value of x in this diagram.

100° $x°$

150° _____

10 Write 0.6 as a simple fraction. _____

11 Simplify $\dfrac{15}{24}$. _____

12 On the number plane, is (3, 0) on the x-axis or the y-axis? _____

13 Draw an isosceles triangle, marking all features.

14 How many faces has a triangular prism?

_____

15 What type of angles are marked between these parallel lines?

_____

PART B: PERCENTAGES AND MONEY   / 25 marks

16 Write each percentage as a simple fraction:

  a 10% _____   **b** $33\dfrac{1}{3}$% _____

17 Convert each percentage to a decimal:

  a 48% _____   **b** 20% _____

18 What is 32% of $180? _____

19 Convert $\dfrac{3}{5}$ to a percentage. _____

20 Convert 1278 cents to dollars. _____

21 Calculate:

  a $3572 + 32.5% × $4200 _____

  b 1500 × 79 cents _____

  c $18.10 × 6 × 1.5 _____

22 1 year = _____ days.

23 1 year = _____ fortnights.

24 1 week = _____ hours.

25 Daniel scored 37 out of 40 in a Geography exam. Convert his mark to a percentage.

_____

26 Convert 0.375 to a percentage. _____

27 A 12-can pack of cola costs $13.95. Calculate the price of 1 can to the nearest cent.

_____

28 Increase $470 by 3%. _____

29 Jay bought a car for $9560. He sold it for $12 000. What was his percentage profit on the cost price, correct to one decimal place?

_____

30 Each week, Huynh saves $15 of her pocket money. How much will she save in 1 year?

_____

31 What percentage of 48 is 20? _____

32 How much simple interest is earned if $14 800 is invested at 6% p.a. for 3 years?

_____

33 Decrease $78 by 10%. _____

34 What percentage is $3.84 of $48? _____

35 By how much is $52 194 greater than $37 000?

_____

36 If 8% of a number is 40, then what is the number? _____

PART C: CHALLENGE   Bonus   / 3 marks

Lisa and Bart start working at a bank, but each person is paid in a different way. Lisa is paid $1 for the first day, $2 for the second day, $4 for the third day, and so on, with her pay doubling each day. Bart is paid $12 every day.

Will Lisa's *total* income ever be higher than Bart's? If so, when?

_____

9780170454537

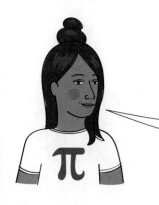

EARNING MONEY CROSSWORD 8

HERE'S ANOTHER CROSSWORD WHERE THE ANSWERS ARE ALREADY GIVEN. WHAT STRATEGY CAN YOU USE TO WORK OUT WHERE THOSE WORDS GO?

The answers to this crossword are listed below in alphabetical order. Arrange them in the correct places on the puzzle.

ALLOWABLE	ANNUAL	COMMISSION	DEDUCTION	DOUBLE	EARNINGS
FIFTY TWO	FINANCIAL	FORTNIGHT	GROSS	HALF	HOLIDAY
INCOME	LEAVE	LOADING	MONTH	NET	OVERTIME
PAYG	PER ANNUM	PERCENTAGE	PIECEWORK	RETAINER	
SALARY	TAX	TAXABLE	WAGE	WEEK	

PUZZLE SHEET   PS

8 TIME AND MONEY CALCULATIONS

THIS REVISION SHEET MAKES ME THINK OF THE EQUATION 'TIME = MONEY.' SORRY.

1 Convert to dollars.

a 1457c _____ b 291c _____

c 5040c _____ d 24 408c _____

2 Write as decimals.

a 14% _____ b 55% _____

c 8% _____ d 80% _____

3 Add together 8 hours, $4\frac{1}{2}$ hours, $6\frac{1}{2}$ hours, 10 hours, $3\frac{1}{2}$ hours and 8 hours.

_____

4 Find:

a 28% of $2733 _____

b 73.5% of $110 _____

c 17.5% of $3581 _____

d 0.88 × $569 _____

e 0.09 × $1084 _____

f $295 + 6.5% × 7148 _____

g 640 × 73c _____

h 280 × $2.27 _____

i 490 × 86.1c _____

5 Round each amount to the nearest cent.

a $143.2506 _____

b $21.8471 _____

c $44.2952 _____

6 By how much does $33 458 exceed $18 200?

_____

7 Multiply each number by 1.5.

a 8 _____ b 3 _____

c 4 _____ d 7 _____

8 Complete:

a 1 year = _____ months

b 1 week = _____ days

c 1 fortnight = _____ days

d 1 working week = _____ days

e $\frac{1}{2}$ year = _____ weeks

f 1 fortnight = _____ weeks

9 What percentage is:

a $65.76 of $274? _____

b $121.66 of $308? _____

c $645.12 of $1075.20? _____

10 Debbie bought 34.8 litres of petrol for $50.12. What was the cost per litre, correct to the nearest cent? _____

11 a Increase $727 by 10% _____

b Increase $3048 by 35% _____

c Decrease $515 by 3% _____

d Decrease $244 by 12.25% _____

12 Round each amount to the nearest dollar.

a $8725.70 _____

b $644.35 _____

c $16 299.39 _____

13 78% of an amount is $453.96. What is the amount? _____

9780170454537

14 Tim earns $28 in pocket money each week. How much does he earn:

a in 6 weeks? _____

b per day? _____

c in a year? _____

15 Shreyashi works $7\frac{1}{2}$ hours each weekday and 4 hours on Saturdays and earns a total of $914.66.

a How many hours does Shreyashi work?

_____

b How much does she earn per hour?

_____

16 By how much does $3491 exceed $1400?

_____

17 How many hours and minutes are there between:

a 9:00 a.m. and 5:30 p.m.? _____

b 8:30 a.m. and 1:00 p.m.? _____

c 5:00 a.m. and 2:00 p.m.? _____

d 7:15 a.m. and 3:45 p.m.? _____

18 110% of an amount is $76.34. What is the amount? _____

19 A games console with a marked price of $360 was reduced to $318.60.

a What was the discount? _____

b What was the percentage discount?

_____

20 Dukhyun had his weekly wage increased by 6% to $786.52. What was his old wage? _____

PERCENTAGES WITHOUT CALCULATORS

50% MEANS 1/2, 25% MEANS 1/4, 10% MEANS 1/10, 33 1/3% MEANS 1/3. CAN YOU DO THESE QUESTIONS WITHOUT USING A CALCULATOR?

WORKSHEET

WS

Calculators are not allowed

1 Convert each percentage to a simplified fraction:

a 70% _____

b 12% _____

c 40% _____

d $33\frac{1}{3}$ % _____

e 5% _____

f 75% _____

2 Convert each percentage to a decimal:

a 18% _____

b 9% _____

c 65% _____

d 20% _____

e $12\frac{1}{3}$ % _____

f 88.3% _____

3 Convert each fraction to a percentage:

a $\frac{3}{5}$ _____

b $\frac{1}{5}$ _____

c $\frac{2}{3}$ _____

d $\frac{9}{10}$ _____

e $\frac{1}{8}$ _____

f $\frac{3}{5}$ _____

4 Find:

a 25% of $260 _____

b 10% of $110 _____

c 60% of $80 _____

d $12\frac{1}{2}$ % of $36 _____

e 5% of $180 _____

f 20% of $748 _____

g 1% of $255 _____

h $33\frac{1}{3}$ % of $420 _____

5 Increase:

a $320 by 5% _____

b $42 by 30% _____

c $120 by $66\frac{2}{3}$ % _____

d $700 by 8% _____

6 Find 18% of 500. Circle the correct answer.

A 90

B 9

C 200

D 2

7 Express as a percentage:

a 11 out of 55 _____

b $18 out of $24 _____

c 40 minutes out of 2 hours _____

d 12 cm out of 1 m _____

e 375 mL out of 3 L _____

f 12 goals out of 15 attempts _____

8 Find:

a 80% of 4 L _____

b 5% of $72 _____

c $33\frac{1}{3}$ % of 2 hours _____

d $2\frac{1}{2}$ % of $300 _____

e 1% of 60 _____

f 75% of 5 kg _____

g 15% of $20 _____

h 110% of $350 _____

9 Find 7% of $850. Circle the correct answer.

    A $955   **B** $59.50

    C $595   **D** $95.50

10 Decrease:

    a $150 by 5% _____

    b $77 by 10% _____

    c $250 by 80% _____

    d $440 by $12\frac{1}{2}$% _____

11 50% of a number is 32. What is the number?

_____

12 Which of the following is 22% of $1400? Circle the correct answer.

    A $74.00   **B** $30.80

    C $740.00   **D** $308.00

13 If 9% of an amount is $27, what is the amount?

_____

14 Express each of the following as a percentage:

    a 12 out of 20 _____

    b $40 out of $160 _____

    c 3 months out of 2 years _____

    d 80 kg out of 0.4 t _____

    e 3 hours out of 1 day _____

    f 20 serves out of 25 attempts _____

15 Find 3% of $450. Circle the correct answer.

    A $15.00   **B** $13.50

    C $7.50   **D** $7.00

16 If 15% of an amount is $45, what is the amount? _____

17 What is the price of a $68 pair of jeans after 10% GST has been added? _____

18 There were 24 boys and 40 girls at the cinema. What percentage of the children were girls?

_____

19 30% of a number is 18. What is the number?

_____

20 What is the price of an $84 tripod after a 25% discount?

_____

21 Aaron earns 4% commission on the sale of cars. How much does he earn for selling a $34 500 car? _____

22 Tegan had a weekly wage of $800 but it increased to $832 this week.

    a What was the increase in Tegan's wage?

    _____

    b What was the percentage increase of her wage? _____

23 Of the students at a school, 7% have red hair. If this is 63 students, how many students are there at the school? _____

24 Dario earns $900 per week but spends $270 of it on rent. What percentage of his weekly earnings is this? _____

25 A computer was bought for $2000 and resold for $1700. Calculate the loss as a percentage of the cost price. _____

8 WAGES AND SALARIES

A WAGE IS PAID PER HOUR, WHILE A SALARY IS BASED ON A YEAR. TO CONVERT A SALARY TO A WEEKLY AMOUNT, DIVIDE BY 52.18.

Note: To convert an annual salary to a weekly amount, divide by 52.18 weeks.

Write all answers to the nearest cent where necessary.

1 Ling earns $23.90 per hour. How much is she paid for working:

a 8 hours? _____

b 15 hours? _____

c 38 hours? _____

2 Klaas is paid $29 per hour to mow lawns. How much will he receive for a job taking $2\frac{1}{2}$ hours?

_____

3 Nadine works 3 days per week, for 8 hours per day. She earns $31.40 an hour. How much will she earn for a week of work?

_____

4 Boris works from 8:00 a.m. to 4:30 p.m., 5 days a week. If he earns $24.30 per hour, how much will he receive for the week?

_____

5 Ali works a 9-day fortnight, $8\frac{1}{2}$ hours a day at a rate of $27.58 per hour. What is his wage for the fortnight?

_____

6 Peter earns a salary of $119 342. How much would he receive if he was paid:

a weekly? _____

b monthly? _____

c fortnightly? _____

7 Madonna earns $20.54 per hour. How much will she earn for working 7 hours each weekday and 5 hours on Saturday?

_____

8 An electrician earns $42.90 per hour. How much will he earn for a job that lasts from 11:30 a.m. to 2:00 p.m.?

_____

9 Complete this table

Employee	Day					Total hours	Hourly rate	Wage
	M	T	W	Th	F			
T. Bagg	7	7	7	7	7		$25.80	
B. Haiv	7	7	5	5	7		$23.12	
X. Kuzmi	3.5	7	8	7	5.5		$27.00	
P. Nutbutta	4	6	8	7	6.5		$21.62	
C. Shore	7	5.5	7	7	–		$33.90	

Hours worked each day

10

Teacher's assistant:	$37.80 per hour
10:00 a.m.–2:30 p.m.	4 days a week

a How many hours are worked each week?

_____

b What is the weekly pay?

_____

11 A ski instructor earns $1995 for a 35-hour week. What is her hourly rate of pay?

_____

12 Julie receives $3501.68 per fortnight. How much does she earn in:

a a year? _____

b a week? _____

c a month? _____

13 Vila earns $24.50 per hour.

a How much will he earn for 8 hours work?

_____

b For how many hours must Vila work to earn over $1600?

_____

14 Ryan is a waiter at a restaurant and is paid $36.94 per hour. Last week he worked the following hours:

- Tuesday   5 p.m.–10 p.m.
- Wednesday   6 p.m.–10:30 p.m.
- Friday   5:30 p.m.–11 p.m.
- Saturday   5 p.m.–12 midnight

Calculate Ryan's total pay for the week.

_____

15 Surinder earns $7135 per month. How much is this amount:

a per year? _____

b per fortnight? _____

16 Convert each pay to a weekly amount.

a $112 000 per year _____

b $6378.94 per month _____

c $230.80 per weekday _____

17 Kristine earns $26.20 per hour and receives $1768.50 each fortnight.

a How many hours does Kristine work in a fortnight? _____

b If she works 9 days per fortnight, how many hours does she work each day?

_____

18 Who earns the most? Circle the correct answer.

A Alyce: $4597.64 per fortnight

B Brett: $112 090 salary

C Claire: $10 244 per month

D Dominic: $2444.70 per week

19 Jack earns $10 500 per month. Jill earns $2287 a week. Who earns more per year, and by how much? _____

_____

20 Megan was offered 2 jobs: one that earned $9200 monthly and another that earned $4400 fortnightly. Which job pays more?

_____

_____

9780170454537

(8) EARNING MONEY

DOUBLE TIME MEANS BEING PAID DOUBLE THE USUAL WAGE PER HOUR, WHILE TIME-AND-A-HALF IS BEING PAID 1.5 AS MUCH.

Name:

Due date:

Parent's signature:

Part A	/ 8 marks
Part B	/ 8 marks
Part C	/ 8 marks
Part D	/ 8 marks
Total	/ 32 marks

Note: To convert an annual salary to a weekly amount, divide by 52.18 weeks.

PART A: MENTAL MATHS

🚫 Calculators are not allowed

1 Evaluate 15×3.

_____

2 Write a prime number between 10 and 20.

_____

3 Find the area of this triangle.

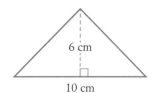

6 cm

10 cm

_____

4 Find 10% of $46.

_____

5 Expand $-2(3a + 9)$.

_____

6 Find $\frac{2}{3}$ of $33. _____

7 Name the marked angle.

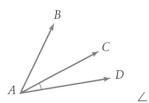

∠ _____

8 Find the mode of 6, 3, 2, 6, 5, 5, 6.

_____

PART B: REVIEW

1 How many fortnights are there in 1 year?

_____

2 Convert $17\frac{1}{2}$% to a decimal.

_____

3 Evaluate:

a $70 931 ÷ 52.18 to the nearest cent

_____

b $17 547 + 0.37 \times ($109 469 - $80 000)$.

_____

4 How many hours and minutes are there between 8.30 a.m. and 4.45 p.m.?

_____

5 Convert 0.015 to a percentage.

_____

6 A trolley bag with a marked price of $84 is discounted by 15%. Calculate its sale price.

_____

_____

7 Evaluate 1840×35 cents.

_____

_____

9780170454537

PART C: PRACTICE

> › Wages, salaries and overtime
> › Commission and annual leave loading

1 Pooja works 8 hours per day for 5 days. Calculate her total pay if she earns $31.50 per hour. _____

2 Pete has an annual salary of $63 082. How much does he earn per week?

_____

3 Kate works 36 hours at $23.84 per hour plus 3 hours overtime at time-and-a-half. Calculate her total earnings.

_____

_____

4 Leo is a real estate agent who sold a house for $435 000. If his commission on the sale was $16 530, what percentage of the house price is this?

_____

5 Gina earns $26.30 per hour for the first 38 hours and double time for any further hours. How much is she paid for working 40 hours?

_____

_____

6 Ahmed earns $3066.90 per fortnight. How much does he earn in a year?

_____

7 Rakitu works from 9 a.m. to 5 p.m. Monday to Friday at $22.50 per hour. Calculate his:

a weekly pay

_____

b annual leave loading if it is 17.5% of 4 weeks' normal pay

_____

PART D: NUMERACY AND LITERACY

1 What is a **salary**? Name one occupation that earns a salary.

_____

_____

2 Explain how to convert a monthly pay of $3850 into a weekly pay.

_____

_____

3 Ritha works the following hours at a chemist.

Thursday	2 p.m. to 9 p.m.
Friday	1 p.m. to 5 p.m.
Saturday	1 p.m. to 6 p.m.
Sunday	2 p.m. to 7 p.m.

She earns $27.31 per hour with time-and-a-half for hours worked after 5 p.m.

a How many hours did Ritha work in total?

_____

_____

b How many overtime hours did she work?

_____

c Calculate her total pay for the 4 days.

_____

4 Jake earns $1249.92 each week for working 38 hours at normal rates and 5 hours at double time.

a Calculate Jake's normal hourly rate.

_____

b Calculate Jake's weekly pay next week if it will increase by 4%.

_____

5 What extra pay is usually given in December?

_____

LET'S GET READY FOR THE STATISTICS TOPIC. DO YOU KNOW HOW TO FIND THE MEAN, MODE AND MEDIAN OF A SET OF DATA?

WORKSHEET

WS

PART A: **BASIC SKILLS**   / 15 marks

1 Evaluate $(-2)^3$. _____

2 From this diagram, find the value of x as a surd.

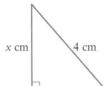

x cm   4 cm

3 cm _____

3 Simplify $\dfrac{12b^2c}{4c}$. _____

4 What is the angle sum of a quadrilateral?

_____

5 Solve $2x + 12 = 5$.

_____

_____

6 If $r = 2.5$ and $h = 5$, evaluate $2\pi r(r + h)$, correct to 2 decimal places.

_____

7 Find the volume of a cube with side length 4 cm. _____

8 Write $66\dfrac{2}{3}\%$ as a simple fraction.

_____

9 Factorise $16y - 64xy$.

_____

10 Write 0.000 25 in scientific notation.

_____

11 Expand $-2(4x - 10)$.

_____

12 Simplify 4.9×10^3.

_____

13 If Kevin's heart beats 72 times per minute, how long will it take to beat 1620 times?

_____

14 Simplify 42 : 36. _____

15 Find the value of θ, correct to the nearest minute.

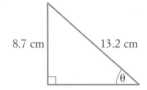

8.7 cm   13.2 cm θ _____

PART B: **STATISTICS**   / 25 marks

16 Name each type of graph shown.

a

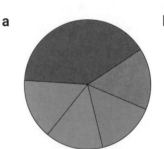

b

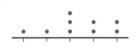

_____

_____

c

8
6
4
2

1 2 3

_____

d

_____

17 For the data 8, 9, 4, 3, 6, 8, 4, 2, find the:

a mean _____

b range _____

18 For the data 2, 4, 6, 5, 4, 6, 4, 2, 6, find the:

a modes _____

b median _____

19 For this data set, find:

4 3 2 2 1 3 2 4 1 2

2 2 2 3 1 2 5 2 3 1

a the mean _____

b the median _____

20 The mean of the 5 values below is 6.

What is the missing value?

4 __ 3 7 6

21 A group of people were surveyed about how many hours they slept last night.

9 8 7 9   6   7 9 10

8 10 7 6 11 10 8 9

7 8 7 10 10   8 9 9

a Complete this frequency table.

Score	Tally	Frequency
6		
7		
8		
9		
10		
11		

b How many people were surveyed? _____

c What was the most popular value? _____

d What is the statistical name for the value in part **c**? _____

e What is the median? _____

f What was the lowest value? _____

g How many people slept for over 8 hours?

_____

h What is the range? _____

i Draw a dot plot for this data.

9780170454537

22 a Complete the frequency table below for these data values:

7 9 3 5 6 10 8   9 10 8

6 5 . 2 7 6   5 9 10   7 7

Score, x	f	fx
2		
3		
4		
5		
6		
7		
8		
9		
10		
Totals:		

b Calculate the mean of the data in the table.

_____

PART C: CHALLENGE   Bonus   / 3 marks

A different executive works on each floor of an office building with 5 floors. Use the 4 clues below to determine whose office is on which floor.

Clue 1: Aspinall works on a floor above Ellsmore and below Badger.

Clue 2: Cooper works on a floor other than the top or bottom.

Clue 3: Davis works on a floor above Badger.

Clue 4: Cooper does not work on a floor adjacent to Davis or Ellsmore.

_____

_____

_____

_____

9 STATISTICS REVIEW

WHEN FINDING THE MEDIAN, DON'T FORGET TO ARRANGE THE VALUES IN ORDER FIRST!

WS WORKSHEET

1 Discrete or continuous data?

a Number of pets owned _____

b Distance from home to school _____

c Shoe size _____

d Running speed _____

2 A group of people was surveyed on the number of people living in their homes.

4   3   4   5   2   4   3   5   7   4

5   6   5   4   3   5   2   3   6   5

Find:

a the mean _____

b the range _____

c the mode _____

d the median _____

e what fraction of homes had 5 people.

_____

3 The scores (out of 10) for a group of dancers are shown in this dot plot.

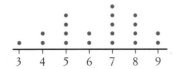

Find:

a the number of dancers _____

b the mean _____

c the mode _____

d the median _____

e the range _____

f what percentage of dancers scored 6 or more.

_____

4 The ages of students in a university class are shown in the stem-and-leaf plot.

Stem	Leaf
1	8 8 9 9 9
2	0 1 2 2 2 4 4 4 5 6 6 8
3	0 0 1 3 6
4	2 8

a How many students are in the class?

_____

b What are the modes? _____

c Find the median. _____

d Find the mean, correct to one decimal place.

_____

e Which value is the outlier? _____

f What fraction of students are in their 20s?

_____

5 A die was rolled a number of times and the results are shown in the frequency table.

Score, x	Frequency, f	fx
1	4	
2	6	
3	7	
4	8	
5	5	
6	6	
Total		

a Complete the fx column. _____

b How many times was the die rolled?

_____

9780170454537

c Calculate the mean, correct to 2 decimal
   places. _____

d Find the median. _____

e Find the mode. _____

f What fraction of rolls landed on a number less
   than 3? _____

g What percentage of rolls landed on an odd
   number? _____

Score, x	Frequency, f	Cumulative frequency
1	33	
2	50	
3	32	
4	22	
5	8	
6	5	

6 The number of people travelling in each car on
a road is shown in the frequency table.

a Complete the cumulative frequency column.

b How many cars were counted? _____

c Calculate the mean. _____

d Find the median. _____

e Find the range. _____

f Find the mode. _____

g Construct a frequency histogram and polygon
   for this data.

h Describe the shape of the distribution.

_____

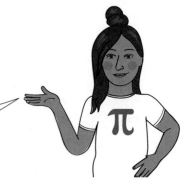

HERE'S ANOTHER CLUE FOR YOU: ANOTHER NAME FOR AVERAGE IS 'MEASURES OF CENTRAL TENDENCY' OR 'MEASURES OF LOCATION'

Clues across

1 The shape of data that is evenly spread about its centre

3 A formal column graph

6 See 26 down

9 A distribution with a tail that points to the right is _____ skewed

12 All of the items being studied or counted

15 The mathematics of collecting and analysing data

20 The middle value in a data set

21 Eye colour is this type of data

22 Mean, median and mode are measures of _____

24 To collect information for statistical purposes

25 The collecting of information about the whole population

26 The shape of data that is not symmetrical

27 $\dfrac{\text{sum of values}}{\text{number of values}}$

28 Type of sample for which any item has an equal chance of being selected

Clues down

2 Type of frequency found by adding up all previous frequencies

4 Part of the population selected for study

5 Bad influence that makes a sample unrepresentative

7 The f in fx

8 Facts or information

10 The tail in a negatively-skewed distribution points this way

11 Numerical data that can be measured on a smooth scale

13 A formal line graph is called a frequency _____

14 Type of numerical data that can only take on separate distinct values

16 Skewed data have a _____ that points to the left or right

17 A simple frequency graph is the dot _____

18 The shape of data with a tail pointing to the left is _____ skewed

19 A value in a data set that is much different from the others

21 A group of data values close together

23 Most popular data value(s)

26 (and 6 across) _____-and-_____ plot

⑨ DATA 1

STATISTICS ARE EVERYWHERE. I LIKE READING THE STATISTICS OF A BASKETBALL MATCH, SUCH AS POINTS PER GAME AND FIELD GOAL PERCENTAGE.

Part A	/ 8 marks
Part B	/ 8 marks
Part C	/ 8 marks
Part D	/ 8 marks
Total	/ 32 marks

PART A: MENTAL MATHS

🔲 Calculators not allowed

1 What solid shape has this net?

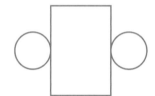

_____

2 Solve $5x - 26 = 3x + 19$.

_____

3 Find 30% of $85. _____

4 Find x.

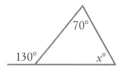

5 If $y = -3x - 1$, find y when $x = 2$.

_____

6 Simplify $\dfrac{5ab^2 \times 10b}{2a^2b}$. _____

7 If p is the probability of a hot day tomorrow, write an expression for the probability that tomorrow will not be hot.

_____

8 Find the volume of this prism.

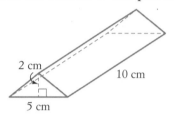

_____

PART B: REVIEW

1 For these data values, find the:

1, 5, 6, 15, 3, 5, 4, 2, 3, 5

a mean _____

b mode _____

c median _____

d range _____

e outlier _____

2 This dot plot shows the number of films watched by a sample of students during the school holidays.

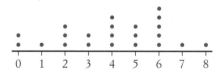

a How many students were surveyed?

_____

b What was the mode?

_____

c What was the median? _____

HW HOMEWORK

PART C: PRACTICE

> › The mean, median, mode and range
> › Histograms and stem-and-leaf plots

1 a (2 marks) Complete this frequency table from a survey on the number of computers owned per home in a street.

Score, x	Frequency, f	fx
3	5	
4	3	
5	7	
6	6	
7	4	
Totals:		

b How many homes were surveyed? _____

c Find the mean. _____

d Find the median. _____

e Draw a frequency histogram for this data.

2 This back-to-back stem-and-leaf plot shows the points scored by 2 junior cricket teams per match during a season.

Piglets		Chicks
3	1	1 4 6
8 0	2	3 5 5 9
7 7 5	3	2 2 8
9 6 3 2 2	4	0 4 5
7 3 1	5	2

a Which team scored better and why?

_____

_____

b Find the median for the Piglets.

_____

c Find the range for the Chicks.

_____

PART D: NUMERACY AND LITERACY

1 Explain the meaning of:

a mode _____

_____

b outlier _____

_____

c histogram _____

_____

2 Which measure is affected most by an outlier: the mean, median or mode?

_____

3 Complete:

a The mean, median and mode are all called measures of _____.

b The symbol $\bar{x}$ stands for _____.

c Where some values in a data set are bunched together is called a

_____.

4 List 5 numbers that have a median of 17 and a mean of 12.

_____

_____

_____

HOMEWORK

HW

⑨ DATA 2

Name:

Due date:

Parent's signature:

Part A	/ 8 marks
Part B	/ 8 marks
Part C	/ 8 marks
Part D	/ 8 marks
Total	/ 32 marks

HOMEWORK

PART A: MENTAL MATHS

🖩 Calculators are not allowed

1 What is the total area of the faces of this cube?

3 cm

_____

2 Solve $3(x + 10) = 4 - 2x$.

_____

3 Find p, giving a reason.

$p°$ $52°$

_____

4 Write 4 numbers that have a mean of 7 and a range of 9. _____

5 (2 marks) Name this shape and state how many axes of symmetry it has.

_____

6 Simplify $\dfrac{3a}{4} + \dfrac{a}{3}$. _____

7 Find the shaded area in terms of π.

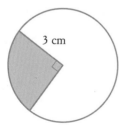

3 cm

_____

PART B: REVIEW

1 For this dot plot, find:

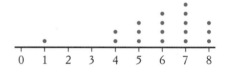

0 1 2 3 4 5 6 7 8

a where the data are clustered

_____

b the mode _____

c the range if the outlier is ignored _____

d the median _____

2 a (2 marks) Complete this frequency table from a survey on the number of children per family.

Score, x	Frequency, f	Cumulative frequency
0	8	
1	18	
2	20	
3	5	
4	1	

b Find the median. _____

c Find the mode. _____

9780170454537

PART C: PRACTICE

> › The shape of a distribution
> › Comparing data sets

1 This stem-and-leaf plot shows the ages of the people at an Anzac Day service.

Stem	Leaf
4	0 7 8
5	1 3 4 4 5 8
6	1 4 6 9
7	3 6 8
8	4 5
9	1

a How many were at the service? _____

b Describe the shape of the data.

_____

c Find the range. _____

d Where are the ages clustered? _____

2 These dot plots show the maximum daily temperatures in 2 cities.

Brisbane

20 21 22 23 24 25 26 27 28 29 30 31 32 33

Melbourne

20 21 22 23 24 25 26 27 28 29 30 31 32 33

Which city:

a had temperatures that were more spread out?

_____

b was warmer? _____

c had the higher mode? _____

d had a median of 26°C? _____

PART D: NUMERACY AND LITERACY

1 a Draw a frequency histogram for the children data from Part B, question **2**.

b Describe the shape of the data.

_____

2 Complete:

a A cluster describes where data is

_____

b A distribution is _____ if the data is evenly spread or balanced about its centre.

c A distribution is _____ skewed if it has a tail that points to the right.

3 Write a word that describes:

a an extreme value _____

b the value with the highest frequency

_____

c a line graph that shows frequencies of numerical data _____

9 DATA 3

WE'RE NEAR THE END OF THE TOPIC NOW. HOW WELL CAN YOU ANALYSE A SET OF DATA? DO YOU KNOW THE DIFFERENCE BETWEEN DISCRETE AND CONTINUOUS DATA? CAN YOU SPOT BIAS IN A SURVEY OR SAMPLE?

Part A	/ 8 marks
Part B	/ 8 marks
Part C	/ 8 marks
Part D	/ 8 marks
Total	/ 32 marks

PART A: MENTAL MATHS

🖩 Calculators are not allowed

1 Convert $\frac{29}{6}$ to a mixed numeral. _____

2 Find p.

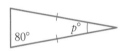

3 A marble is selected from a bag of 2 red, 4 blue and 6 orange marbles. What is the probability that it is not blue? _____

4 Simplify $3xy^2 + 4x^2y - xy^2 + 10x^2y$.

_____

5 Find 40% of $760. _____

6 Find the area of this trapezium.

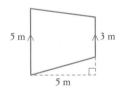

_____

7 What is the length of the hypotenuse in a right-angled triangle if the other 2 sides are 8 m and 6 m? _____

8 Expand $(x - 9)(3x + 5)$.

_____

PART B: REVIEW

1 For this dot plot:

```
                        •               •   •
    •       •           •       •   •   •   •
    •   •   •   •       •   •   •   •   •   •
    1   2   3   4   5   6   7
```

a find the mode _____

b describe the shape of the distribution

_____

c the range _____

d the median _____

2 This back-to-back stem-and-leaf plot shows the marks scored by a class in a maths test.

Boys		Girls
5 2 0	4	8
8 4 2	5	0 7 9
7 6 2 1	6	3 5 7 8
8 3 1	7	5 5 6 9
6 0	8	1 2 9

a Which gender:

i scored better? _____

ii had the higher range? _____

iii had the higher median? _____

b What percentage (correct to one decimal place) of students scored over 80?

_____

PART C: PRACTICE

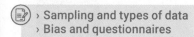
› Sampling and types of data
› Bias and questionnaires

1 Determine whether a census or a sample should be used in each investigation.

a Predicting who will win the next election.

_____

b Finding the number of teachers working in Catholic schools.

_____

c Finding the percentage of boys born at a local hospital last year.

_____

2 Classify whether each type of data is categorical or numerical. If numerical, then state whether it is continuous or discrete.

a School report grades: A, B, C, D, E.

_____

b The number of part-time workers.

_____

c The speed of a truck.

_____

d Brand of tablet device.

_____

3 Visitors to a news website were asked to vote on whether the Prime Minister is doing a good job. Explain why this sample of voters is biased.

_____

_____

_____

PART D: NUMERACY AND LITERACY

1 Which measure of central tendency:

a can have more than one value?

_____

b is also called 'the average'?

_____

c is the sum of values divided by the number of values?

_____

2 Describe the shape of a negatively-skewed distribution.

_____

_____

3 Write the word that means:

a how often each value appears in a data set

_____

b a survey of the whole population

_____

c data that can be measured on a smooth scale with no gaps

_____

d an unwanted influence on a sample or questionnaire that makes it unrepresentative of all items or views

_____

9 MEAN, MEDIAN AND MODE

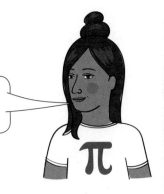

REMEMBER: A SET OF DATA CAN HAVE MORE THAN ONE MODE.

For each set of data, find the mean (to 2 decimal places where appropriate), median, mode(s) and range.

Set		Mean	Median	Mode(s)	Range
Set 1	8, 11, 11, 15, 17, 20, 21				
Set 2	10, 10, 16, 16, 20, 25, 26, 28				
Set 3	3, 4, 5, 8, 11, 11, 14, 14, 14, 19, 20				
Set 4	0, 1, 1, 1, 4, 5, 8, 10, 12				
Set 5	18, 19, 20, 20, 20, 25, 27, 31, 34, 40				
Set 6	7, 15, 19, 25, 29, 31, 40, 40, 50, 55				
Set 7	54, 54, 60, 64, 64, 64, 68, 68, 70				
Set 8	2, 1, 5, 3, 4, 5, 2				
Set 9	91, 84, 86, 88, 90, 83				
Set 10	10, 20, 30, 40, 50				
Set 11	5, 4, 6, 3, 2, 4, 6, 9, 4, 7, 3, 2, 3				
Set 12	16, 17, 19, 15, 17, 19, 14, 16, 17, 20				
Set 13	7, 6, 9, 9, 8, 7, 7, 3				
Set 14	17, 20, 19, 22, 21, 17, 100				
Set 15	8, 10, 12, 7, 8, 10, 9, 8, 10, 8, 11				
Set 16	21, 23, 20, 22, 21, 23, 24, 25				
Set 17	47, 51, 48, 50, 48, 52, 51				
Set 18	9, 9, 9, 9, 9, 10				
Set 19	1, 0, 2, 3, 0, 1, 4, 2, 3, 0, 1, 1, 5, 4, 3, 2				
Set 20	5, 9, 4, 6, 7, 8, 6, 5, 3, 2, 6, 4, 8				

9780170454537

BEFORE WE START THE SURFACE AREA AND VOLUME TOPIC, LET'S REVISE SOME MEASUREMENT FACTS, INCLUDING PERIMETERS AND AREAS OF SHAPES.

PART A: BASIC SKILLS   / 15 marks

1 Write the coordinates of P.

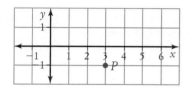

2 Find the mode of:

8, 5, 3, 2, 3, 5, 4, 5. _____

3 Expand $(x - 5)^2$. _____

4 List the 4 factors of 15. _____

5 Evaluate 2^6. _____

6 If $r = -3$, evaluate $2 - 2r$. _____

7 Find the value of x in this triangle, correct to one decimal place.

_____

8 What type of angles are a and b shown below?

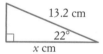

_____

9 Solve $3(x - 4) = 21$. _____

_____

10 $3 : 5 = 15 :$ _____

11 What percentage is $18.75 of $156.25?

_____

12 Simplify $2p^2 \times 3p^4$. _____

13 Find the value of m in this diagram.

_____

14 Evaluate $\dfrac{3}{4} + \dfrac{1}{5}$. _____

15 Petrol costs 135.9 cents per litre. How many litres were bought for $33.98? _____

PART B: MEASUREMENT   / 25 marks

16 1 tonne = _____ kg

17 1 hour = _____ seconds

18

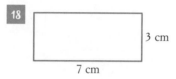

For this rectangle, find:

a the perimeter _____

b the area. _____

19 Find the area of this triangle.

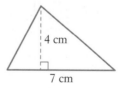

_____

20 What is:

4 hours 48 minutes + 2 hours 22 minutes?

_____

21 How many axes of symmetry has a rhombus?

_____

22 Find the area of this parallelogram.

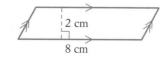

_____

23 If the area of a square is 12.25 cm², what is the length of one side? _____

24 For this circle, find correct to 2 decimal places:

a its circumference

_____

b its area

_____

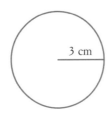
3 cm

25 For this figure, find:

a the perimeter

_____

b the area.

_____

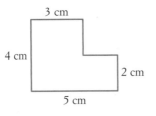
3 cm
4 cm
2 cm
5 cm

26 For this right-angled triangle, find:

a the value of *h*

_____

b the perimeter

_____

c the area.

_____

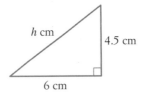
h cm
4.5 cm
6 cm

27 a Name this solid.

_____

b What is its volume? _____

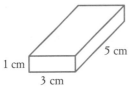
5 cm
1 cm
3 cm

c How many faces has this solid?

_____

28 What fraction of this circle is shaded?

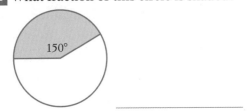
150°

_____

29 What is the time in New York when it is 8:30 p.m. in Sydney if New York is 15 hours behind Sydney? _____

30 a Name this solid.

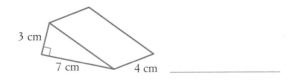
3 cm
7 cm
4 cm

_____

b How many faces has the solid?

_____

c What is the shape of each face?

_____

_____

31 Draw a quadrilateral, *ABCD*, where *AB* ∥ *CD*. What type of quadrilateral is *ABCD*?

_____

PART C: CHALLENGE   Bonus   / 3 marks

A rectangular yard measuring 18 metres by 8 metres must be fenced. If fence posts will be placed 2 metres apart, how many posts will be needed?

HOW WELL DO YOU KNOW YOUR AREA FORMULAS FOR TRIANGLES, QUADRILATERALS AND CIRCLES? THIS WORKSHEET WILL HELP YOU PRACTISE USING THEM.

Name each shape, write its area formula and calculate its area, as shown in question **1**.

1
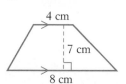

Trapezium

$A = \dfrac{1}{2}(a + b)h$

$A = 42 \text{ cm}^2$

2
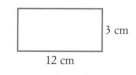

Rectangle

$A = lw$

_____

3
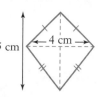

Kite

$A = \dfrac{1}{2}xy$

_____

4
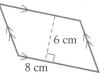

_____
_____
_____

5
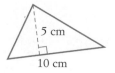

6

7

8

9
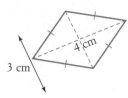

10

11

12
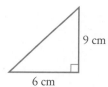

13
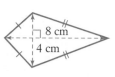

14

15

16
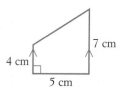

17
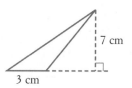

18
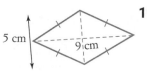

19

20

21
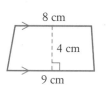

22

23
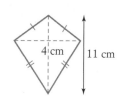

24

10 A PAGE OF PRISMS AND CYLINDERS

IT'S UP TO YOUR TEACHER TO DECIDE WHETHER YOU SHOULD FIND THE SURFACE AREA OR VOLUME OF EACH SOLID SHAPE BELOW.

Teacher's tickbox: For each shape, find (correct to one decimal place for cylinders):

❏ its surface area ❏ its volume

1

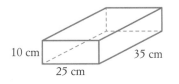

10 cm

25 cm

35 cm

2

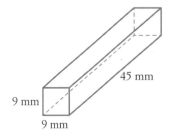

45 mm

9 mm

9 mm

3

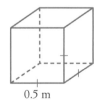

0.5 m

4

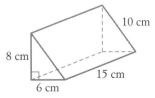

10 cm

8 cm

6 cm

15 cm

5

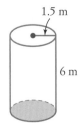

1.5 m

6 m

6

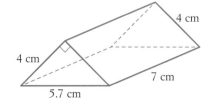

4 cm

4 cm

5.7 cm

7 cm

7

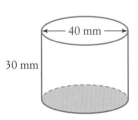

40 mm

30 mm

8

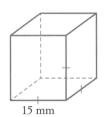

15 mm

9

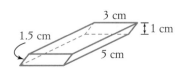

3 cm

1 cm

1.5 cm

5 cm

10

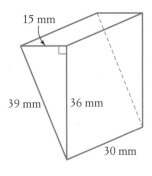

15 mm

39 mm

36 mm

30 mm

11

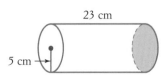

23 cm

5 cm

12

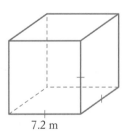

7.2 m

HERE ARE THE WORDS FROM THIS TOPIC, HERE IS THE CROSSWORD PUZZLE. GO TO IT!

The answers to this crossword puzzle are listed below in alphabetical order. Arrange them in the correct places.

ACCURACY	AREA	BASE
BREADTH	CAPACITY	CENTURY
CIRCLE	CIRCUMFERENCE	COMPOSITE
DIAGONAL	DIAMETER	DIMENSIONS
ERROR	ESTIMATE	FACES
HECTARE	HOUR	KITE
LENGTH	LIMITS	LITRE
MASS	METRIC	PERIMETER
PRISM	RADIUS	RECTANGLE
RHOMBUS	SECTOR	SQUARE
SURFACE	TONNE	TRAPEZIUM
TRIANGLE	VOLUME	

(10) AREA 1

LET'S PRACTISE OUR AREA FORMULAS. DO YOU KNOW WHAT EACH VARIABLE IN THE FORMULA STANDS FOR? MAKE SURE YOU DON'T CONFUSE PERIMETER AND AREA.

PART A: MENTAL MATHS

🖩 **Calculators not allowed**

1 Convert 19:35 to 12-hour time. _____

2 What is the probability of rolling a factor of 6 on a die?

_____

3 Write Pythagoras' theorem for this triangle.

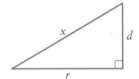

_____

4 Factorise $10x^2y - 15xy^2$. _____

5 Simplify $28 : 16$. _____

6 Simplify 9^{-2}. _____

7 Write a algebraic expression for the number of days in w weeks. _____

8 Solve $6x - 26 = x + 1$.

_____

_____

_____

_____

PART B: REVIEW

1 Match each area formula to its correct shape: circle, parallelogram, rhombus, square, trapezium, triangle

a $A = \frac{1}{2}(a+b)h$ _____

b $A = \frac{1}{2}bh$ _____

c $A = \frac{1}{2}xy$ _____

d $A = \pi r^2$ _____

e $A = bh$ _____

2 Find the perimeter of each shape.

a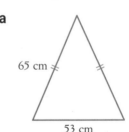

65 cm

53 cm

b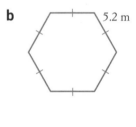

5.2 m

_____ _____

3 Find the shaded area.

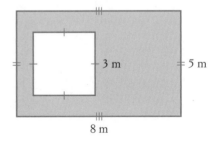

3 m

5 m

8 m

_____

_____

9780170454537

PART C: PRACTICE

> › Perimeters and areas of composite shapes
> › Areas of triangles and quadrilaterals

1 For this shape, find:

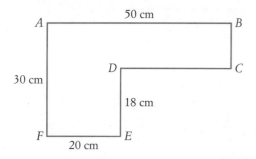

a the perimeter

_____

b the area

_____

2 Find the area of each shape.

a

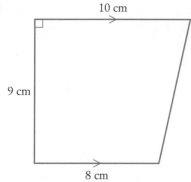

_____

b

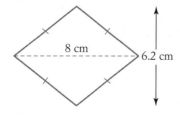

_____

c

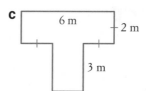

_____

d

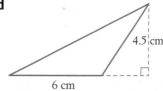

_____

e

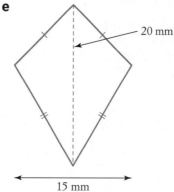

_____

f

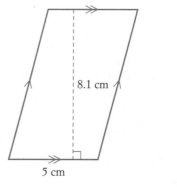

_____

PART D: NUMERACY AND LITERACY

1 a Write 'cm²' in words.

_____

b Describe how big a 'cm²' is.

_____

_____

c Complete: 1 cm² = _____ mm²

2 What word means the distance around a shape, the total length of its sides?

_____

3 The value of a square's perimeter in metres is the same as the value of its area in m².

What is the length of each side of the square?

_____

_____

4 For this rectangle, write a simplified algebraic expression for:

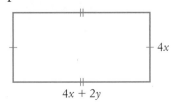

a its perimeter

_____

_____

b its area

_____

_____

5 Find, correct to one decimal place, the perimeter of the trapezium in question **2 a** in Part C.

_____

_____

_____

_____

9780170454537

Name:

Due date:

Parent's signature:

Part A	/ 8 marks
Part B	/ 8 marks
Part C	/ 8 marks
Part D	/ 8 marks
Total	/ 32 marks

I DON'T CONFUSE THE FORMULAS FOR THE CIRCUMFERENCE AND AREA OF A CIRCLE. I REMEMBER THAT CIRCUMFERENCE IS A LENGTH SO IT INVOLVES R, BUT AREA HAS SQUARE UNITS SO IT INVOLVES R².

PART A: MENTAL MATHS

Calculators are not allowed

1 a Draw a triangular prism.

b How many faces has a triangular prism?

2 Evaluate 37×101. _____

3 Write $\frac{2}{3}$, 0.65, 0.6 and 68% in descending order.

4 Expand $(x + 6)(x - 10)$.

5 Find p.

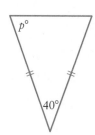

6 How many hours and minutes are there between 4.45 p.m. and 10.15 p.m.?

7 If a cube has side length 2 m, what is its volume?

PART B: REVIEW

1 a Find, as a surd, the length of the hypotenuse in this triangle.

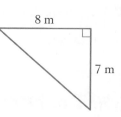

8 m

7 m

b Find, correct to 2 decimal places, the perimeter of the triangle above.

2 Find the area of each shape.

a

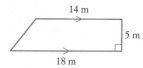

14 m

5 m

18 m

b

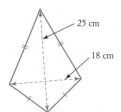

25 cm

18 cm

c

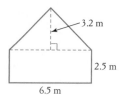

3.2 m

2.5 m

6.5 m

d

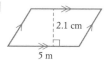

2.1 cm

5 m

HOMEWORK

HW

3 Find the perimeter of this shape.

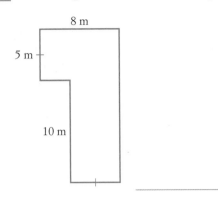

8 m

5 m

10 m

4 Write the formula for the area of a circle with radius *r*. _____

PART C: *PRACTICE*

> › Circumferences and areas of circular shapes

1 Find, correct to 2 decimal places, the circumference of each circle.

a

8 cm

b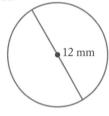

12 mm

_____

_____

2 Find, correct to 2 decimal places, the area of each circle in question **1**.

a _____

b _____

3 Find, correct to 3 significant figures, the perimeter of each sector.

a

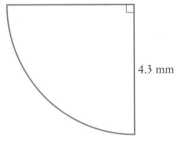

4.3 mm

_____

_____

b

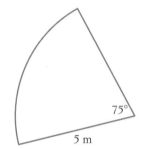

75°

5 m

_____

_____

4 Find, correct to 3 significant figures, the area of each sector in question **3**.

a _____

b _____

9780170454537

PART D: NUMERACY AND LITERACY

1 Why is π called an irrational number?

_____

_____

2 Find, correct to 2 decimal places, the area of each shape.

a

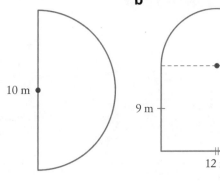

10 m

b

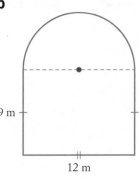

9 m

12 m

_____   _____

_____   _____

_____   _____

3 What word means:

a the perimeter of a circle?

_____

b the distance from one edge of a circle to the opposite edge, through the centre?

_____

c the shape of a 'pizza slice' of a circle?

_____

4 Find, correct to one decimal place, the area of each shaded region.

a

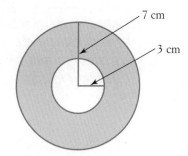

7 cm

3 cm

_____

_____

_____

b

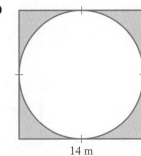

14 m

_____

_____

_____

⑩ SURFACE AREA

TO FIND THE SURFACE AREA OF ANY SOLID, FIND THE AREA OF EACH FACE OR SURFACE, THEN ADD THEM TOGETHER.

Part A	/ 8 marks
Part B	/ 8 marks
Part C	/ 8 marks
Part D	/ 8 marks
Total	/ 32 marks

PART A: MENTAL MATHS

🚫 Calculators are not allowed

1 Evaluate:

   a 5.728×100 _____

   b $30 - 6 \times 4 + 20$ _____

2 Find x, giving a reason.

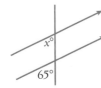

_____

3 What word means the most common value(s) in a set of data?

_____

4 Complete: $63 : 72 = 21 :$ _____

5 Solve $3(3x - 7) = -12$.

_____

_____

6 Simplify $4ab + ab^2 - 7ab^2 - 10ab$.

_____

_____

7 If Ravi and Cindy share a prize of $2000 in the ratio 2 : 3, what is Ravi's share?

_____

_____

PART B: REVIEW

1 Find, correct to 2 decimal places, the area of each shape.

a

10 m

_____

_____

_____

b

16 cm

_____

_____

_____

c

8 mm

6 mm

_____

_____

d

14 m
120°
14 m

_____
_____
_____

2 Find correct to 2 decimal places the perimeter of each shape in question **1**.

a _____
_____

b _____
_____

c _____
_____

d _____
_____

PART C: PRACTICE

📝 › Surface areas of prisms and cylinders

1 Find the surface area of each solid (correct to 2 decimal places for cylinders).

a

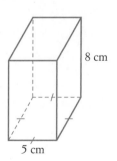

8 cm
5 cm

_____
_____

b

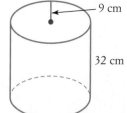

9 cm
32 cm

_____
_____

c

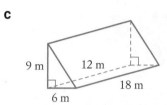

9 m   12 m
6 m   18 m

_____
_____

d

8.5 cm

13.4 cm

_____
_____

e

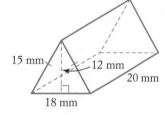

15 mm   12 mm   20 mm
18 mm

_____
_____

f

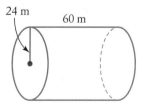

24 m   60 m

_____
_____

HOMEWORK

2 (2 marks) Find the surface area of this prism.

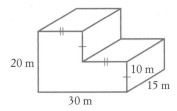

_____

_____

PART D: *NUMERACY AND LITERACY*

1 a How many faces has a cylinder? _____

b What are the shapes of those faces?

_____

_____

2 Write the formula for the curved surface area of a cylinder with radius r and height h.

_____

3 This cube has an open top. Find its (external) surface area.

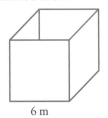

6 m

_____

_____

4 Describe what the surface area of a solid means.

_____

_____

_____

5 This cylinder has open ends. Find, correct to 3 significant figures, its surface area.

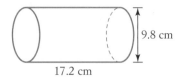

9.8 cm

17.2 cm

_____

6 A cube has a surface area of 54 cm². What is the length of each side?

_____

_____

9780170454537

Name:

Due date:

Parent's signature:

Part A	/ 8 marks
Part B	/ 8 marks
Part C	/ 8 marks
Part D	/ 8 marks
Total	/ 32 marks

DON'T FORGET: SQUARE UNITS SUCH AS M² FOR SURFACE AREA, BUT CUBIC UNITS SUCH AS CM³ FOR VOLUME.

HW HOMEWORK

PART A: *MENTAL MATHS*

🚫 Calculators are not allowed

1 Simplify $5p^3 \times 8pr$. _____

2 Convert $\dfrac{40}{6}$ to a mixed numeral. _____

3 If $x = -7$, evaluate $8x - 9$. _____

4 Does the point $(0, -4)$ lie on the x-axis or y-axis? _____

5 Find y.

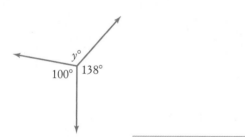

_____

6 Simplify $\dfrac{6a}{7} - \dfrac{5a}{6}$. _____

7 For this set of data, find:

Stem	Leaf
1	1 4 4 5
2	0 3 6
3	2 4
4	7

a the median _____

b the range _____

PART B: *REVIEW*

1 Find, correct to 2 decimal places, the area of each shape.

a

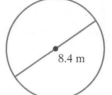

8.4 m

_____

b

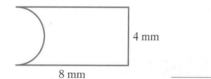

4 mm

8 mm

_____

c

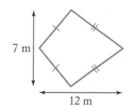

7 m

12 m

_____

d

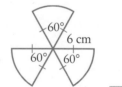

60°
6 cm
60° 60°

_____

2 Find the surface area of each solid (correct to 2 decimal places for **c** and **d**).

a

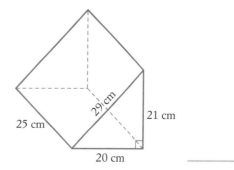

29 cm

25 cm

21 cm

20 cm

_____

b

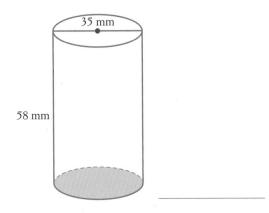

2.3 mm
6.4 mm
4.1 mm

_____

c

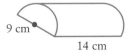

35 mm

58 mm

_____

d

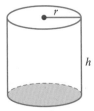

9 cm

14 cm

_____

PART C: **PRACTICE**

> Volumes of prisms and cylinders

1 Write down the formula for the:

r

h

a volume of a cylinder

_____

b surface area of a cylinder

_____

2 Find the volume of each solid (correct to 2 decimal places for **a** and **c**).

a 6.4 m

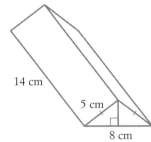

14.7 m

_____
_____

b

14 cm

5 cm

8 cm

_____
_____

c

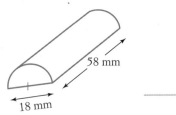

58 mm

18 mm

_____
_____

d

6.2 m
8.7 m
5.8 m
10.3 m
9.7 m

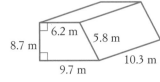

_____
_____

3 **a** What does the formula $V = Ah$ describe?

_____

b What do A and h represent in the formula?

_____
_____

9780170454537

PART D: NUMERACY AND LITERACY

1 What shape is the cross-section of a cylinder?

_____

2 This can has an open top.

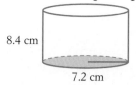

8.4 cm

7.2 cm

a Find its capacity in litres, correct to 2 decimal places.

_____

b Find its surface area, correct to the nearest square centimetre.

_____

_____

3 a Write 'm³' in words.

_____

b Describe how big a 'm³' is.

_____

_____

c Complete 1 m³ = _____ cm³

4 Find, correct to the nearest cm³, the volume of each solid.

a

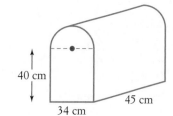

30 cm

25 cm

15 cm

_____

_____

_____

_____

b

40 cm

34 cm

45 cm

_____

_____

_____

_____

(11) STARTUP ASSIGNMENT 11

> THIS TOPIC, COORDINATE GEOMETRY AND GRAPHS, SHOWS THAT PATTERNS IN
> ALGEBRA CAN ALSO BE REPRESENTED AS PATTERNS ON A NUMBER PLANE.
> WORK THROUGH THESE QUESTIONS TO BE PREPARED FOR THIS TOPIC.

WORKSHEET · WS

PART A: BASIC SKILLS / 15 marks

1 Calculate 18^7:

a to 3 significant figures _____

b in scientific notation to 3 significant figures.

_____

2 Expand $-2(3x + 5)$. _____

3 True or false? The opposite sides of a parallelogram are equal. _____

4 What is the angle sum of a parallelogram? _____

5 What fraction is 750 m of 5 km? _____

6 Solve $5x + 9 = 2x - 3$.

_____

_____

7 Find the surface area of this prism.

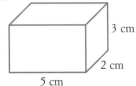

3 cm

2 cm

5 cm

_____

8 Find n if $2^n = \dfrac{1}{2}$. _____

9 Write $\dfrac{7}{12}$ as a decimal. _____

10 Decrease $130 by 15%. _____

11 A cyclist travels at 11 km/h. How far will she travel in $2\dfrac{1}{2}$ hours? _____

12 How many axes of symmetry has a parallelogram? _____

13 Calculate Tarren's weekly wage if she earns $28.24 per hour and works Monday to Friday 7 a.m. to 4 p.m.

_____

14 Divide $184 in the ratio 5 : 3.

_____

_____

PART B: THE NUMBER PLANE

/ 25 marks

15 Evaluate:

a $\dfrac{2-5}{15-6}$ _____

b $\dfrac{-8-4}{5-3}$ _____

c $\sqrt{(8-2)^2 + (13-5)^2}$ _____

d $\sqrt{(-4+8)^2 + (-1-2)^2}$ _____

16 What is the average of 18 and 12? _____

17 Write an expression for the average of a and b.

_____

18 a What are the coordinates of P on the number plane below?

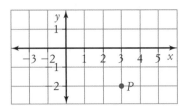

_____

b What is the distance between P and $(3, 1)$?

_____

9780170454537

19 Find d:

3 cm

5 cm d cm

a as a surd _____

b correct to 2 decimal places. _____

20 If $m = \dfrac{b - a}{d - c}$, $a = 3, b = 7, c = -1$ and $d = 1$, evaluate m. _____

21 Complete each table of values.

a $y = 4x - 2$

x	−1	0	1	2	3
y					

b $y = 10 - 2x$

x	−1	0	1	2	3
y					

22 Write a formula for this table of values.

$y = $ _____

x	−1	0	1	2	3
y	0	3	6	9	12

23 a If $\triangle PQR$ is right-angled at R, write the coordinates of R.

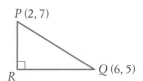

$P\,(2, 7)$

R $Q\,(6, 5)$ _____

b What is the length of RQ? _____

c Use Pythagoras' theorem to find the length of PQ as a surd.

_____

_____

How can you get exactly 6 litres of water from a river using only 2 buckets: one which holds 4 litres and one which holds 9 litres?

9 L

4 L

WS   WORKSHEET

⑪ DRAWING GRADIENTS

GRADIENT MEANS STEEPNESS OR SLOPE, SO ON THIS WORKSHEET WE'RE PRACTISING DRAWING LINES WITH DIFFERENT GRADIENTS. GRADIENT = RISE/RUN.

For each question, use the grid paper to draw a line with the given gradient.

1 Gradient = 3

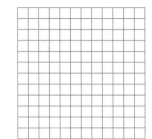

2 Gradient = 2

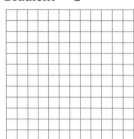

3 Gradient = 5

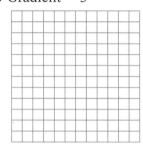

4 Gradient = −2

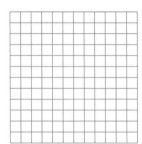

5 Gradient = −1

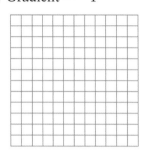

6 Gradient = 1

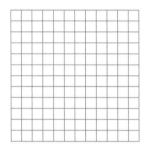

7 Gradient = $\frac{1}{2}$

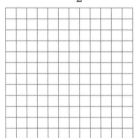

8 Gradient = 0

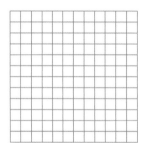

9 Gradient = −3

10 Gradient = $\frac{1}{4}$

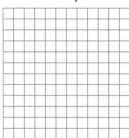

11 Gradient = 6

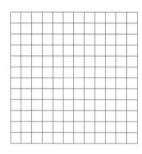

12 Gradient = −4

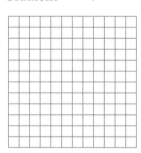

13 Gradient = $-\frac{1}{3}$

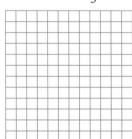

14 Gradient = −8

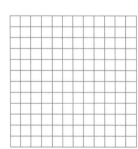

15 Gradient = $\frac{3}{4}$

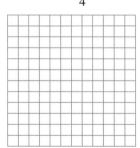

9780170454537

YOUR TEACHER WILL DECIDE WHAT NEEDS TO BE DONE WITH THESE INTERVALS ON THE NUMBER PLANE, WHETHER IT BE FINDING THEIR LENGTHS, MIDPOINTS OR GRADIENTS.

Teacher's tickbox

For each interval:

❏ calculate its length (correct to 2 decimal places)

❏ find the coordinates of its midpoint

❏ calculate its gradient

❏ find the *y*-intercept of the line that runs through the interval

❏ find the equation of the line that runs through the interval.

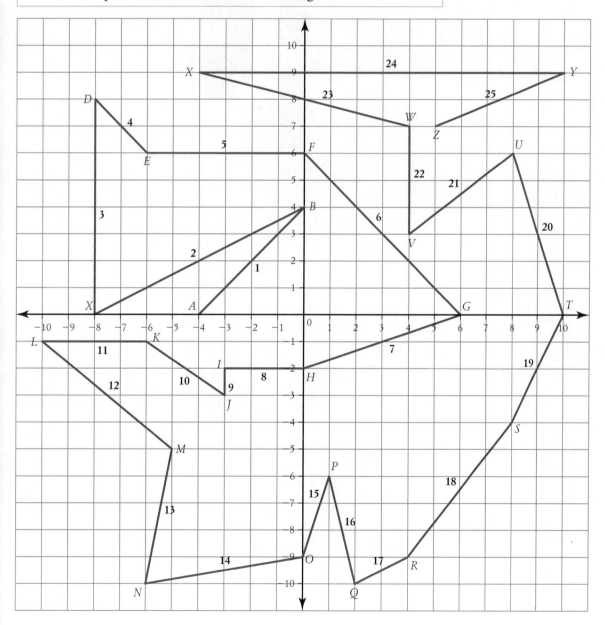

Mixed length answers: 6, 8.94, 8.25, 2.24, 4.47, 5, 8, 5.66, 3, 3.61, 5.39, 8.49, 6.40, 4, 6.32, 6.40, 2.83, 6.08, 4, 3.14, 4.12, 1, 5.10, 3.16, 14

11 COORDINATE GEOMETRY CROSSWORD

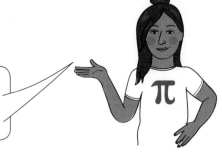

HERE'S A DIFFERENT TYPE OF CROSSWORD ... THE ANSWERS ARE GIVEN, BUT THEIR LETTERS ARE JUMBLED! THEY'RE ALL COORDINATE GEOMETRY WORDS THOUGH.

Unscramble the word clues and place them in the crossword.

Across

3 RPOAABLA

4 PHRAG

8 CLERIC

9 XESA

10 TTNCOSNA

11 VUCER

12 URN

13 INAQUOTE

14 QUITCARAD

16 ILEARN

19 RECENTIPT

20 SIXA

23 MYSTERYM

24 NEIL

26 ITDANCES

Down

1 TIGERDNA

2 ALZITHONOR

3 NOPOORTRIP

5 SIRE

6 REHABPLOY

7 VIRALNET

11 ECNOVCA

15 SIDRAU

17 NODIMPIT

18 VXTREE

21 PESTSSEEN

22 RUDS

25 CATLIVER

PUZZLE SHEET

PS

11 COORDINATE GEOMETRY 1

WE'RE PRACTISING FINDING THE LENGTH, MIDPOINT AND GRADIENT OF AN INTERVAL ON THE NUMBER PLANE. LENGTH USES PYTHAGORAS' THEOREM, MIDPOINT USES AVERAGES.

Name:

Due date:

Parent's signature:

Part A	/ 8 marks
Part B	/ 8 marks
Part C	/ 8 marks
Part D	/ 8 marks
Total	/ 32 marks

HOMEWORK

PART A: MENTAL MATHS

🚫 Calculators are not allowed

1 Simplify 40 : 36. _____

2 Find the sum of $9.90, $1.40, $2.50 and $8.50.

_____

3 Find the area of this triangle.

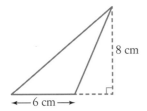

8 cm

← 6 cm →

_____

4 If a die is rolled 80 times, how many times would you expect a 5 or 6 to come up?

_____

5 Evaluate 0.0705×100. _____

6 Write 0.009 51 in scientific notation.

_____

7 For these data values, find:

3, 6, 6, 7, 8, 10, 11, 16

a the mean _____

b the range _____

PART B: REVIEW

1 Simplify $\dfrac{15}{25}$. _____

2 Complete this table for $y = 2x - 1$.

x	−2	−1	0	1	2
y					

3 Find the average of:

a −2 and 8 _____

b 1 and 4 _____

4 If $y = x + 4$, find y when:

a $x = 0$ _____

b $x = 3$ _____

c $x = -7$ _____

5 Find p.

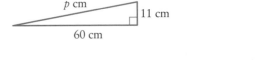

p cm

11 cm

60 cm

_____

_____

9780170454537

PART C: PRACTICE

> › Length and midpoint of an interval
> › Gradient of a line

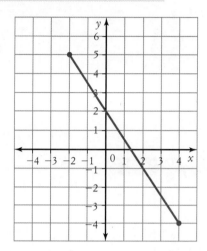

1 For this interval, find:

  a its length as a surd

  _____

  _____

  b its midpoint

  _____

  c its gradient

  _____

2 Draw a line with a gradient of 1.

3 The interval AB has endpoints $A(0, 6)$ and $B(4, 8)$.

  a Write the coordinates of C if $\triangle ABC$ is right-angled at C.

  _____

  b Find the length of AB, correct to one decimal place.

  _____

  c Find the midpoint of AB.

  _____

d Find the gradient of AB.

_____

PART D: NUMERACY AND LITERACY

1 The interval PQ has endpoints $P(-4, 4)$ and $Q(8, -6)$.

  a What does the midpoint of an interval mean?

  _____

  _____

  b Explain how to find the midpoint of PQ.

  _____

  _____

  c Sketch PQ on a number plane and explain how you know that its gradient is negative.

  _____

  _____

  _____

  _____

  d (2 marks) $\text{Gradient} = \dfrac{\text{rise}}{\text{run}}$. What do rise and run mean? Write the rise and run for PQ.

  _____

  _____

  _____

2 Which famous rule is being used when we calculate the length of an interval?

_____

3 (2 marks) Show that the line going through points $(3, 4)$ and $(7, 4)$ has a gradient of 0. Why does it have a gradient of 0?

_____

_____

_____

11 COORDINATE GEOMETRY 2

HAVE YOU NOTICED THAT YOU CAN TELL THE GRADIENT AND Y-INTERCEPT OF A LINE FROM ITS EQUATION?

$y = 3x - 2$ HAS A GRADIENT OF 3 AND A Y-INTERCEPT OF -2.

Name:

Due date:

Parent's signature:

Part A	/ 8 marks
Part B	/ 8 marks
Part C	/ 8 marks
Part D	/ 8 marks
Total	/ 32 marks

PART A: MENTAL MATHS

🚫 Calculators not allowed

1 Evaluate:

a $-9 \times (-8)$ _____

b $\dfrac{2}{3} - \dfrac{1}{4}$ _____

c $\left(\dfrac{4}{5}\right)^2$ _____

d $15 \times 4 \times 7$ _____

2 What decimal is shown by the dot on scale?

_____

3 Decrease $180 by 15%.

_____

4 Find x.

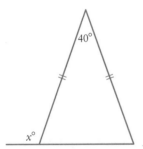

_____

5 Find the mode of these values.

5, 2, 1, 0, 2, 3, 2, 3, 4, 5, 3

_____

PART B: REVIEW

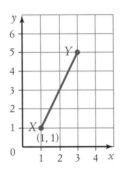

1 For interval XY, find:

a its length, correct to one decimal place.

_____

b its midpoint.

_____

c its gradient.

_____

2 Find the gradient of each line.

a

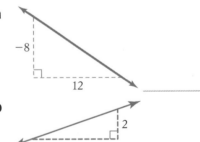

b

3 The interval PQ has endpoints $P(10, -6)$ and $Q(-6, 4)$. Find:

a the length of PQ as a surd. _____

b the midpoint of AB. _____

c the gradient of AB. _____

9780170454537

PART C: PRACTICE

> › Graphing linear equations
> › Solving linear equations graphically
> › Direct proportion

1 **a** Complete the table for $y = \dfrac{1}{2}x + 2$.

x	-2	-1	0	1	2
y					

b Graph $y = \dfrac{1}{2}x + 2$ on a number plane

c Write down the y-intercept of this line.

_____

d Find the gradient of the line.

_____

2 P is directly proportional to n.

When $n = 16$, $P = 40$. Find P when $n = 30$.

_____

_____

3 **a** Graph $y = 5 - 2x$ on a number plane.

b Test whether $(5, -4)$ lies on the line.

_____

c Use the graph to solve $5 - 2x = 2$.

_____

PART D: NUMERACY AND LITERACY

1 **a** Sketch a line with gradient -1 that goes through the origin.

b Write down the equation of this line.

_____

2 What name is given to the value at which a line crosses the x-axis?

_____

3 Describe the graph of the line $y = 2$.

_____

_____

4 In $y = kx$, which variable is the constant of proportionality?

_____

5 The drop, $D°C$, in temperature of a cooling body varies directly with time, t minutes.

a If the temperature dropped 4°C in 5 minutes, write a formula for D in terms of t.

_____

_____

b How long will it take the body to drop 10°C?

_____

_____

6 How is the gradient of an interval calculated?

_____

_____

(11) COORDINATE GEOMETRY 3

IN THIS ASSIGNMENT, WE'LL BE LOOKING AT THE GRAPHS OF NON-LINEAR EQUATIONS. WHY DO YOU THINK THEY ARE CALLED 'NON-LINEAR'?

Name:

Due date:

Parent's signature:

Part A	/ 8 marks
Part B	/ 8 marks
Part C	/ 8 marks
Part D	/ 8 marks
Total	/ 32 marks

PART A: MENTAL MATHS

🔲 Calculators not allowed

1 Find the area of this square.

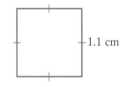

1.1 cm

_____

2 If the probability that you will live past age 100 is 0.09, what is the probability that you won't live past 100? _____

3 Dinali pays 30% of her weekly wage in tax. If she pays $276 tax, what is her wage?

_____

4 Simplify $x \times y + x \times 6$. _____

5 Find y.

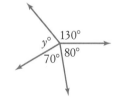

130°
$y°$
70° 80°

_____

6 Evaluate $-8 + 14 + (-2)$. _____

7 Write the formula for the area of a kite.

_____

8 Find the median of these values:

12, 5, 12, 18, 11, 20, 17, 3.

_____

PART B: REVIEW

1 a Graph the line $y = 2x - 3$.

b Find its y-intercept. _____

c Find its x-intercept. _____

d Find its gradient. _____

2 The mass, M grams, of a chemical substance varies directly with its volume, V cm³.

a Find a formula for M if 60 cm³ of the chemical has a mass of 150 g. _____

b Calculate the mass of 604 cm³ of the chemical.

_____

c Calculate the volume of 1 kg of the chemical.

_____

3 Find the equation of this line.

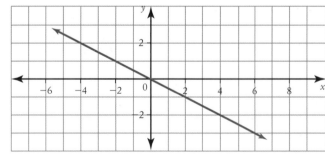

9780170454537

PART C: PRACTICE

> › Graphing parabolas
> › Graphing circles

1 a Complete the table for $y = \frac{1}{2}x^2 + 1$.

x	-2	-1	0	1	2
y					

b Graph $y = \frac{1}{2}x^2 + 1$ on a number plane

c Write down the y-intercept of this curve.

_____

d What is the special name for this curve?

_____

e Is this curve concave up or concave down?

_____

2 a (2 marks) Describe the graph of $x^2 + y^2 = 9$.

_____

b Sketch the graph of $x^2 + y^2 = 9$.

PART D: NUMERACY AND LITERACY

1 a What are the coordinates of the turning point of the graph in Part C, question **1**?

_____

b What is another name for this turning point?

_____

2 What is the equation of a circle with centre $(0, 0)$ and diameter 10?

_____

3 Find the equation of each line.

a

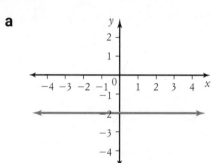

b

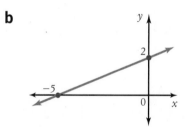

4 Find the gradient of the line with equation
$$y = 5 - 3x.$$

_____

5 What does r represent in the equation
$x^2 + y^2 = r^2$? _____

6 What does c represent in the equation
$y = mx + c$? _____

(11) COORDINATE GEOMETRY REVISION

WE'RE UP TO THE END OF THE TOPIC NOW. COMPLETE THESE QUESTIONS, AND YOU WILL HAVE MASTERED COORDINATE GEOMETRY AND GRAPHS.

Name:

Due date:

Parent's signature:

Part A	/ 8 marks
Part B	/ 8 marks
Part C	/ 8 marks
Part D	/ 8 marks
Total	/ 32 marks

PART A: MENTAL MATHS

🚫 Calculators not allowed

1 Evaluate:

   a $420 \div 6$ _____

   b $10^4 \times 10^5 \div 100$ _____

2 Find, correct to 3 significant figures, the area of this circle.

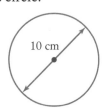

10 cm

_____

3 What is the probability that the sun will rise in the west tomorrow? _____

4 Expand and simplify $3(4 - 2x) + 10x$.

_____

5 Increase $800 by 75%. _____

6 Find the median of these values.

Stem	Leaf
0	2 3 3 8
1	0 4
2	1 9

_____

7 Find the perimeter of this shape.

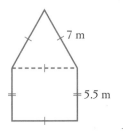

7 m

5.5 m

_____

PART B: REVIEW

1 **a** Complete the table for $y = -x^2 + 3$.

x	-2	-1	0	1	2
y					

   b Graph $y = -x^2 + 3$ on a number plane.

   c Estimate the x-intercepts of this curve.

_____

   d What is the vertex for this curve?

_____

   e Is this curve concave up or concave down?

_____

2 (2 marks) Label the graphs $y = \frac{1}{2}x^2$, $y = \frac{1}{3}x^2$ and $y = x^2$.

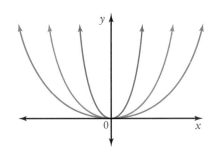

3 Find the equation of this circle.

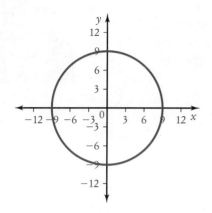

4 Q is directly proportional to t.

If $Q = 54$ when $t = 7.5$, find Q when $t = 23$.

_____

_____

_____

_____

PART D: NUMERACY AND LITERACY

1 Estimate the gradient of this line.

PART C: PRACTICE

> Coordinate geometry revision

1 Test whether (3, 11) lies on the line with equation $y = 3x + 2$.

_____

_____

2 The interval AB has endpoints $A(2, -2)$ and $B(-4, 6)$. Find:

a the midpoint of AB _____

_____

b the length of AB _____

_____

c the gradient of AB _____

3 a (2 marks) Write the gradient and y-intercept of the line $y = -x - 2$.

_____

b Graph the line $y = -x - 2$.

2 Complete:

a The graph of $y = -4x^2 + 1$ is called a

_____ and is concave _____ .

b The variable for gradient is the letter

_____ .

c For an interval joining 2 points, finding the average of their x-coordinates and the average of their y-coordinates will give the

_____ of the interval.

d The gradient of an interval is calculated by dividing the _____ by the _____ .

3 Write an example of:

a a quadratic equation _____

b a linear equation _____

4 What are the coordinates of the vertex of the graph of $y = -3x^2$? _____

(12) STARTUP ASSIGNMENT 12

LET'S GET READY FOR PROBABILITY,
THE TOPIC ABOUT CHANCE. IT'S USED IN PREDICTING
WHAT THE WEATHER WILL BE LIKE, HOW A SPORTS TEAM
WILL PERFORM, WHAT WILL HAPPEN TO THE ECONOMY.

PART A: **BASIC SKILLS** / 15 marks

1 Calculate, correct to the nearest dollar, the value of a \$27 000 car after 2 years if it depreciates at 11% p.a. _____

2 Describe the **mode** of a list of data values.

_____

_____

3 What is the scale on a map if 4 cm represents 200 m? _____

4 Find the average of b, $b + 10$ and $b - 1$.

_____

_____

5 What is the size of $\angle NOP$?

6 Find n if $(3a^n)^3 = 27a^9$.

_____

7 Find the value of x in this diagram.

8 For this sector, calculate to 2 decimal places:

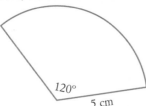

a its perimeter _____

b its area _____

9 Solve $3(x + 7) = 5x + 14$.

_____

_____

_____

10 Is the line with equation $y = -2$ horizontal or vertical? _____

11 For $X(-2, 3)$ and $Y(-10, -12)$ on the number plane, find the length of XY.

_____

_____

_____

12 Calculate Yasmin's pay if she earns \$24.70 per hour for 8 hours, plus $2\frac{1}{2}$ hours at time-and-a-half.

_____

_____

_____

13 For the line with equation $y = \frac{1}{2}x + 3$, find:

a its gradient _____

b its y-intercept. _____

PART B: **PROBABILITY** / 25 marks

14 If the chance that it rains on Saturday is 36%, what is the chance it doesn't rain? _____

15 What is the probability that a number drawn at random from the numbers 1 to 20 is:

a less than 9? _____

b a square number? _____

c divisible by 3? _____

16 A ticket is drawn at random from a box containing 5 blue, 7 white, 3 red and 3 yellow tickets. Find the probability this ticket is:

a red _____

b blue or white _____

c not red _____

17 What is the probability a person chosen at random has a birthday in a month beginning with J? _____

18 Sort these events and place them in order, from most to least likely:

D Rolling an odd number on a die

L Buying your lunch at the canteen tomorrow

Y The next student walking past being in Year 12

P A letter in the post for you today

I You watch a YouTube video today

S You're at school before 8:30 a.m. tomorrow

_____

19 Evaluate:

a $1 - 0.45$ _____

b $1 - \dfrac{5}{6}$ _____

20 What does it mean if an event has a probability of 1? _____

21 Write, as a decimal, the probability of a '50-50 chance'. _____

22 What are the possible outcomes for the result of a soccer match between the Reds and the Blues?

_____

23 Henry, Isabella, Jack, Kathy and Lisa wrote their names on separate cards. What is the probability that a card chosen at random has a boy's name? _____

24 There are 400 tickets in a raffle. If Jessie buys 8 tickets, find the decimal probability she wins first prize. _____

25 If the probability of having 2 boys in a 3-child family is $\dfrac{3}{8}$, what is the probability of not having 2 boys? _____

26 A tossed coin came up heads 18 times and tails 22 times. What percentage of tosses showed tails? _____

27 A card is drawn at random from a deck of playing cards. What is the probability of selecting a card showing:

a diamonds? _____

b an ace? _____

c a jack, queen or king? _____

d an even number? _____

28 What percentage (to one decimal place) of the alphabet are vowels? _____

PART C: CHALLENGE Bonus / 3 marks

Jane, Megan, Robert and Steve sit together in a row for a group photo. How many possible seating arrangements are there?

(12) MATCHING PROBABILITIES

YOU CAN FIND THE MIXED ANSWERS TO THESE QUESTIONS NEXT PAGE. MATCH UP THE NUMBER AND LETTER TO DECODE THE MESSAGE THERE.

Match each event with its correct probability in the box next page. The capital letters in the box form a code that gives the answer to the question below it.

1 Being born on a Tuesday

2 A 30% chance of rain tomorrow

3 Selecting a black card from a normal deck of cards

4 Selecting an Ace from a deck of cards

5 Selecting a card that is not a heart

6 Selecting a picture card (J, K, Q)

7 Choosing one of the 2 cash prizes behind these cards:

8 In a 2-child family, both children being boys

9 A horse has a $\frac{10}{55}$ chance of winning

10 Rolling a total of 5 on a pair of dice

11 Rolling a total of 11 on a pair of dice

12 The Sharks have a 65% chance of winning their next game

13 Selecting a red ball from a bag of 12 red balls, 5 yellow balls and 3 blue balls

14 Selecting a consonant from the letters of the words NEW CENTURY

15 Winning a prize in a 500-ticket raffle when there are 4 prizes

16 Winning a TV on this wheel:

17 Selecting a female delegate from a student council of 16 girls and 12 boys

18 Being in a plane crash

19 In 3 tosses of a coin, getting exactly 2 heads

20 Randomly typing in a 4-digit PIN correctly

21 There are 12 chances out of 98 of selecting an 'E' in a game of Scrabble

22 Getting green on a traffic light when the green-amber-red time ratio is 100 s: 2 s: 48 s

23 Getting a bullseye if the arrow lands inside this target

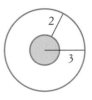

24 Getting the letters C-A-T in order when the letters A, C and T are drawn out of a bag randomly

25 Being in a house fire

26 Ben wins money 15 times out of 100 in scratch lotteries

27 A medical test has a 0.05 chance of giving an
incorrect result

28 Selecting a phone number ending in 0

29 Getting tails and a 6 when tossing a coin and
die together

30 Not selecting the black sock from a bag of 2 red
socks, 3 white socks and 1 black sock

T $\frac{1}{4}$	N $\frac{1}{6}$	F $\frac{2}{7}$	G $\frac{3}{8}$	I $\frac{1}{2}$	P $\frac{1}{7}$	O $\frac{2}{9}$
A $\frac{3}{10}$	Y $\frac{1}{2}$	W $\frac{1}{10}$	R $\frac{2}{11}$	E $\frac{1}{12}$	Z $\frac{1}{13}$	N $\frac{3}{13}$
E $\frac{1}{18}$	X $\frac{3}{20}$	B $\frac{1}{20}$	U $\frac{4}{25}$	Q $\frac{6}{49}$	U $\frac{1}{125}$	
V $\frac{1}{800}$	H $\frac{1}{10\,000}$	A $\frac{1}{20\,000}$	S $\frac{2}{3}$	M $\frac{3}{4}$	D $\frac{3}{5}$	
L $\frac{4}{7}$	J $\frac{7}{10}$	C $\frac{5}{6}$	K $\frac{13}{20}$			

What does the phrase Buckley's chance mean?

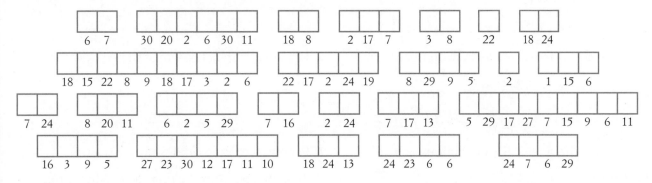

(12) TREE DIAGRAMS

MANY STUDENTS HAVE TROUBLE WITH TREE DIAGRAMS, BUT THEY SIMPLY SHOW ALL THE POSSIBLE ALTERNATIVE TIMELINES OF HOW A SITUATION CAN TURN OUT!

This probability tree diagram represents the possible outcomes from 4 tosses of a coin.

1st toss	2nd toss	3rd toss	4th toss

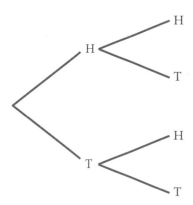

1 a How many branches are there on the tree diagram for the first toss?

_____

b What is the probability of obtaining a head from one toss?

_____

2 a How many branches are there on the tree diagram for the second toss?

_____

b What is the probability of obtaining 2 heads from 2 tosses?

_____

3 How many branches would there be for the third toss?

_____

4 Follow the pattern and continue the tree diagram above for the third toss. What is the probability of obtaining 3 heads from 3 tosses?

_____

5 In 3 tosses, what is the probability of:

a 2 heads and 1 tail (the order doesn't matter)?

_____

b 3 tails? _____

c 1 head and 2 tails? _____

6 Continue the tree diagram to show the outcomes of the fourth toss. How many branches are there?

_____

7 What is the probability of obtaining 4 heads from 4 tosses?

_____

8 In 4 tosses, what is the probability of:

   a 2 heads and 2 tails? _____

   b 1 head and 3 tails? _____

   c 3 heads and 1 tail? _____

9 Use your tree diagram to list all 16 possible outcomes when a coin is tossed 4 times (for example, HHHH, HHHT, . . .)

_____

_____

_____

(12) PROBABILITY 1

PROBABILITY CAN BE WRITTEN AS A FRACTION OR DECIMAL FROM 0 TO 1, OR A PERCENTAGE FROM 0% TO 100%.

Name:

Due date:

Parent's signature:

Part A	/ 8 marks
Part B	/ 8 marks
Part C	/ 8 marks
Part D	/ 8 marks
Total	/ 32 marks

HW HOMEWORK

PART A: *MENTAL MATHS*

[calculator icon] Calculators not allowed

1 The probability of rain today is 0.32. What is the probability that it won't rain? _____

2 Find the mean of these values:

12, 11, 22, 8, 9. _____

3 Simplify $\dfrac{14p^4y}{56p^2y^3}$. _____

4 Write tan B for this triangle.

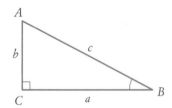

_____

5 What type of line has a gradient of 0?

_____

6 Evaluate $-1 \times 4 + (7 - 9)$. _____

7 Find the sale price of a $50 item after a 15% discount.

_____

8 Find the area of this parallelogram.

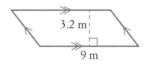

_____

PART B: *REVIEW*

1 Write the sample space for:

a tossing a coin

_____

b rolling a die

_____

c the suit of a playing card

_____

2 Express $\dfrac{4}{5}$ as a decimal. _____

3 Express $\dfrac{23}{60}$ as a percentage. _____

4 Use one of the following terms to describe each event: impossible, very unlikely, unlikely, even chance, likely, very likely, certain.

a An even number is rolled on a die.

_____

b It will snow at your school tomorrow.

_____

c You will have breakfast tomorrow.

_____

9780170454537

PART C: PRACTICE

> › Probability
> › Relative frequency
> › Venn diagrams

1 a What is the probability of rolling 4 on a die?

_____

b Describe the complementary event to rolling 4 on a die.

_____

_____

c What is the sum of the probabilities of an event and its complementary event?

_____

2 This Venn diagram shows the results of a survey on whether people preferred watching movies at the cinema or at home.

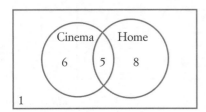

a How many people prefer:

i both the cinema and home? _____

ii the cinema only? _____

b What is the probability that a person chosen at random from the survey prefers either the cinema or home but not both?

_____

3 a A fake coin was tossed and came up heads 21 times and tails 29 times. Write, as a decimal, the relative frequency of tossing tails. _____

b If this coin was tossed 240 times, what is the expected number of times tails will come up?

_____

9780170454537

PART D: NUMERACY AND LITERACY

1 A number is chosen at random from 1 to 20. What is the probability that it is at least 5?

_____

2 Write the meaning of:

a sample space

_____

b random

_____

3 (2 marks) If H stands for the event 'a tossed coin showing heads', what does $P(\bar{H})$ mean?

_____

_____

4 When playing darts, Sam has a 67% chance of hitting the bullseye. What is the probability that she misses the bullseye?

_____

5 What word means the result of a chance experiment or situation?

_____

6 A card is selected at random from a deck of 52 playing cards. What is the probability that it is not a number card? _____

12 PROBABILITY 2

Name:

Due date:

Parent's signature:

Part A	/ 8 marks
Part B	/ 8 marks
Part C	/ 8 marks
Part D	/ 8 marks
Total	/ 32 marks

PART A: MENTAL MATHS

🚫 Calculators not allowed

1 Write the recurring decimal 2.146 464 6 ... using dot notation. _____

2 Complete: $2 : 3 = 14 :$ _____ .

3 What type of line has a negative gradient?

_____

4 Write cos B for this triangle.

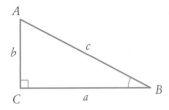

_____

5 Factorise $16xy^2 - 8ky$.

_____

6 What is the time 14 hours 20 minutes after 3.35 p.m.? _____

7 Do the diagonals of a rectangle bisect each other at right angles? _____

8 Find x.

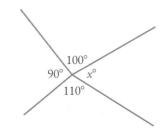

_____

PART B: REVIEW

1 If the probability that the Tigers will win the match is $\frac{5}{7}$, what is the probability that they won't win? _____

2 (2 marks) A pet survey of 18 students showed that 14 students had dogs, 8 students had cats and 1 had neither. Complete the Venn diagram.

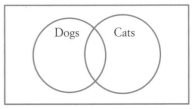

3 a Describe the complementary event to rolling a number greater than 4 on a die.

_____

b What is the probability of this complementary event? _____

4 This Venn diagram shows the number of students who scored an A in their Maths and Science tests.

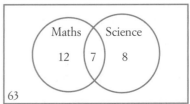

Find the probability that a student selected at random:

a scored an A in Maths or Science _____

b scored an A in Maths or Science, but not both

_____

5 From a group of 80 people, how many would you expect to have a birthday in January?

_____

PART C: PRACTICE

> Two-way tables
> Two-step experiments

1 This table describes students at a school.

	Male	Female
Left-handed	13	19
Right-handed	62	83

How many students are:

a male and left-handed? _____

b male or left-handed? _____

2 a Write the sample space when 2 coins are tossed. _____

b Find the probability of tossing:

　i 2 tails _____

　ii one tail _____

　iii at most one tail _____

3 A sample of shoppers were surveyed on whether they ate breakfast and lunch that day.

	Lunch	No lunch
Breakfast	27	10
No breakfast	45	3

What is the probability that a shopper in the sample:

a ate lunch but not breakfast? _____

b ate one of the meals? _____

PART D: NUMERACY AND LITERACY

1 A coin and die are tossed together.

a (2 marks) Complete this table of outcomes.

		Die					
		1	2	3	4	5	6
Coin	**H**						
	T						

b Find the probability of tossing:

　i a head and 4 _____

　ii a tail and a number less than 3

_____

2 On a Venn diagram, what does the rectangle represent?

_____

3 A survey of sets of traffic lights at 10 a.m. showed 36 greens, 25 reds and 3 ambers. What is the relative frequency of a traffic light showing green?

_____

4 Give an example of a two-step experiment.

_____

_____

5 The digits 1, 3, 7 and 8 are written on separate cards. List all possible 2-digit numbers that can be made from the cards.

_____

_____

⑫ PROBABILITY REVISION

LET'S END THE TOPIC BY PRACTISING OUR TREE DIAGRAMS AND VENN DIAGRAMS. THIS COMES UP A LOT IN PROBABILITY PROBLEMS SO MAKE SURE YOU GET IT RIGHT!

Name:

Due date:

Parent's signature:

Part A	/ 8 marks
Part B	/ 8 marks
Part C	/ 8 marks
Part D	/ 8 marks
Total	/ 32 marks

PART A: MENTAL MATHS

🚫 Calculators not allowed

1 What do x and y stand for the formula for the area of a rhombus $A = \frac{1}{2}xy$?

_____

2 Simplify:

a $(-3a^5)^2$ _____

b $4x^0 - (4x)^0$ _____

3 Find x.

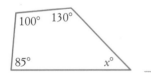

_____

4 Evaluate:

a 20% of $65 _____

b $\frac{2}{3} - \frac{1}{4}$ _____

5 Round 3.0286 to 3 significant figures.

_____

6 Find the mode of this data set. _____

PART B: REVIEW

1 A bag contains one black marble, 5 green marbles, 6 yellow marbles and 3 red marbles.

a If a marble is selected at random from the bag, find the probability that it is:

i red _____

ii black or green _____

iii not yellow _____

b If 100 selections were made from the bag (with replacement), how many times should a green marble come up?

_____

2 This weekend, on each day it is equally likely to rain (R) or not rain ($\bar{R}$).

a Draw a tree diagram for the possible outcomes for Saturday and Sunday this weekend.

b What is the probability that it will:

i not rain on both days? _____

ii rain on Saturday only? _____

iii rain on at least one of the days?

_____

9780170454537

PART C: PRACTICE

📝 › Probability revision

1 Find the probability that a card drawn randomly from a deck of playing cards is:

a a king or queen _____

b an even number _____

2 Two coins were tossed repeatedly.

Outcome	HH	HT	TH	TT
Frequency	30	27	34	26

Find, correct to 3 decimal places, the relative frequency of tossing:

a no heads _____

b at most one head _____

3 This Venn diagram shows which radio station a sample of people listen to.

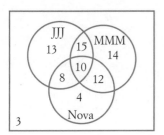

Write how many people from this sample:

a listen to both JJJ and Nova _____

b do not listen to MMM _____

c listen to MMM or Nova _____

d listen to JJJ only _____

PART D: NUMERACY AND LITERACY

1 A group of teachers was surveyed on whether they worked full-time or part-time.

	Full-time	**Part-time**
Male	55	5
Female	64	36

Find the percentage probability that a teacher from this group:

a works part-time _____

b is male and works part-time _____

c is female or works full-time _____

2 Two dice are rolled and the difference between the larger and smaller numbers is calculated.

a (2 marks) Complete this table of outcomes.

−	1	2	3	4	5	6
1						
2						
3						
4						
5						
6						

(1st die across top, 2nd die down side)

b Find the probability of rolling a difference:

i of 5 _____

ii that is odd _____

iii of at most 1 _____

HW HOMEWORK

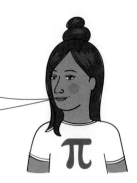

HERE'S ANOTHER CROSSWORD WHERE THE ANSWERS ARE GIVEN. GOOD LUCK FINDING WHERE THEY GO.

The answers to this crossword puzzle are listed below in alphabetical order. Arrange them in the correct places in the puzzle.

PUZZLE SHEET

PS

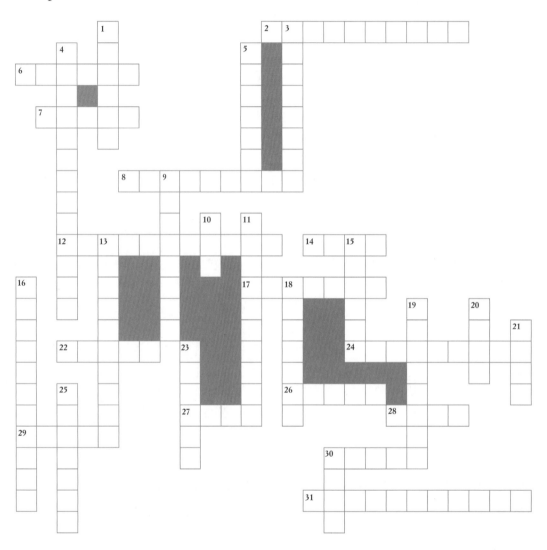

CERTAIN	CHANCE	COIN	COMPLEMENTARY
DIAGRAM	DICE	DIE	EVENT
EXCLUSIVE	EXPECTED	EXPERIMENT	FAVOURABLE
FREQUENCY	IMPOSSIBLE	LEAST	LIST
MOST	MUTUALLY	OUTCOME	PROBABILITY
RANDOM	RELATIVE	REPLACEMENT	SAMPLE
SPACE	TABLE	THEORETICAL	TREE
TRIAL	TWO STEP	TWO WAY	VENN

STARTUP ASSIGNMENT 13 (13)

NOTICED SOME RED QUESTIONS HERE FOR THE CONGRUENT AND SIMILAR FIGURES TOPIC. THESE QUESTIONS ARE EXTRA CHALLENGING AND WILL REQUIRE MORE TIME. SO WILL THE CHALLENGE PROBLEM!

PART A: BASIC SKILLS   / 15 marks

1 What fraction is halfway between $\frac{1}{6}$ and $\frac{1}{4}$?

_____

2 Evaluate $\dfrac{6.5 \times 10^5}{2.5 \times 10^7}$ in scientific notation.

_____

3 If Gavith's salary is $110 596.80 and he pays 30.4% tax, calculate his weekly net wage.

_____

4 Find the size of an exterior angle in an equilateral triangle. _____

5 Simplify $3w^{-2}$. _____

6 If $T = a + (n - 1)d$, find T if $a = -7$, $n = 5$ and $d = \dfrac{1}{2}$. _____

7 Simplify $5m - 3 - 2(3m - 5)$. _____

8 Find the value of h in this diagram, correct to 2 decimal places.

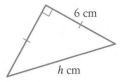

_____

9 Find the median of 11, 6, 1, 15, 14, 1. _____

10 Solve $7k = 5k + 14$. _____

11 Write an algebraic expression for the size of $\angle CBE$ in this diagram.

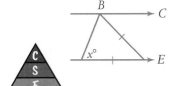

_____

12 Find, correct to 3 significant figures, the perimeter of this semi-circle.

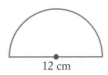

12 cm

_____

_____

13

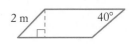

3 cm d cm

4 cm   4 cm

For this triangular prism, find:

a the value of d _____

b the prism's surface area.

_____

_____

14 Find, correct to 2 decimal places, the height of this parallelogram.

2 m   40°

_____

_____

PART B: RATIOS AND SCALE DRAWINGS   / 25 marks

15 Simplify in ratio form:

a 1 cm to 5 m _____

b 40 m to 1 km _____

c 2.5 cm to 1 km _____

16 Solve:

a $\dfrac{x}{5} = \dfrac{15}{25}$ _____

b $\dfrac{h}{12} = \dfrac{18}{108}$ _____

c $\dfrac{130}{400} = \dfrac{t}{20}$ _____

17 A map's scale is 1 : 200 000.

a How far apart are 2 towns if they are 7.5 cm apart on the map? _____

b How long will a 20 km freeway be on the map? _____

18 In a photo, a man 1.8 m tall is 4 cm high. What is the photo's scale? _____

19 A town's main street is 3 km long but on a map it appears 6 cm long. What is the map's scale?

_____

20 Find the value of x in each triangle.

a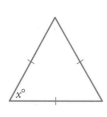

_____

b

_____

c

_____

d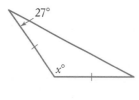

_____

e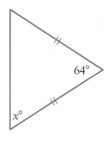

_____

21 Which transformation has been performed on the figure **R** each time?

a R ⟶ Я _____

b R ⟶ ꓤ _____

22 List the 4 tests for congruent triangles.

_____

23 **a** By what scale factor has △RST been enlarged to make △WXY? _____

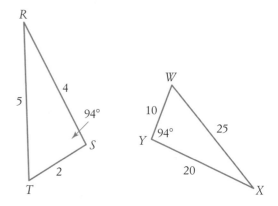

b Which angle in △WXY corresponds to ∠T in △RST? _____

24 **a** Construct this triangle.

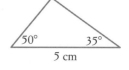

b How long is the shortest side of the triangle?

_____

PART C: CHALLENGE   Bonus   / 3 marks

Find 4 numbers W, X, Y and Z that add to 180, and $W + 2 = X - 2 = Y \times 2 = Z \div 2$.

9780170454537

A PAGE OF CONGRUENT AND SIMILAR FIGURES (13)

I HOPE YOU HAVE YOUR RULER, BECAUSE YOU'LL NEED IT FOR THIS MEASURING AND MATCHING ACTIVITY.

Teacher's tickbox

❏ Match all pairs of congruent and similar triangles below.

❏ Find the scale factor between similar triangles. Express answers as $\dfrac{\text{large}}{\text{small}}$.

(13) FINDING SIDES IN SIMILAR FIGURES

THESE ARE PAIRS OF SIMILAR FIGURES, SO THE 2ND SHAPE IS AN ENLARGEMENT OR REDUCTION OF THE 1ST SHAPE. SAME SHAPE, DIFFERENT SIZE!

WORKSHEET
WS

Teacher's tickbox

For each pair of similar figures:

❏ find the scale factor

❏ find the value(s) of each variable

1

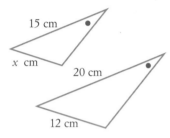

2

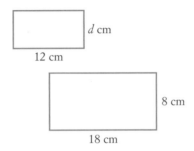

3

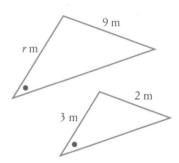

4

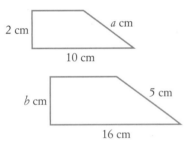

5

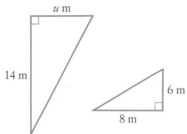

6

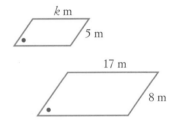

7

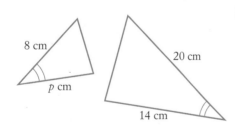

8

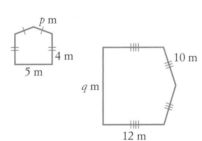

9

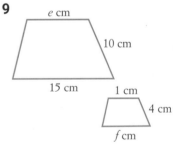

9780170454537

10

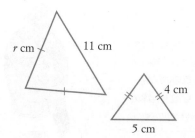

11

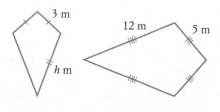

12

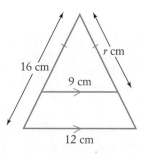

13

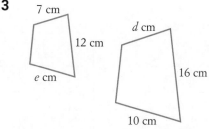

14

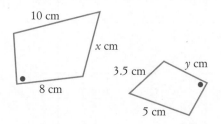

15

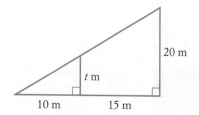

(13) CONGRUENCE AND SIMILARITY CROSSWORD

GEOMETRY TOPICS ALWAYS HAVE A LOT OF JARGON, SO IT'S NO DIFFERENT FOR CONGRUENT AND SIMILAR FIGURES. UNSCRAMBLE THE KEYWORDS NEXT PAGE TO COMPLETE THIS CROSSWORD.

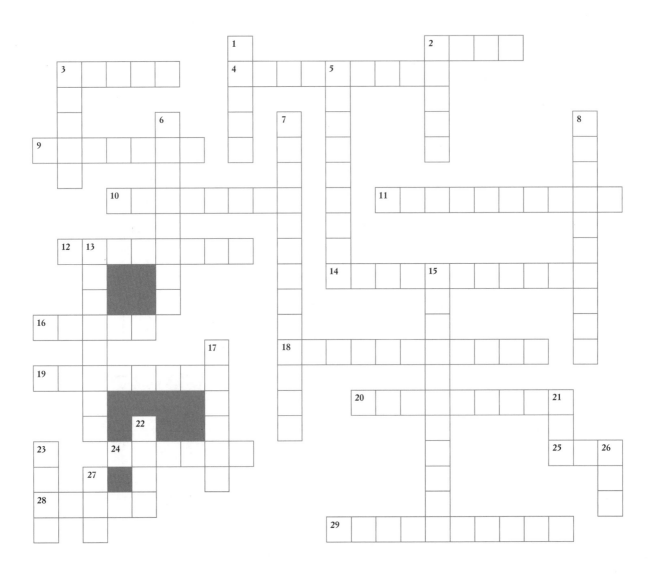

Unscramble the word clues and place them in the crossword.

Across

2 REAA

3 UERLR

4 RUENOCTDI

9 SRIAMLI

10 DCUIENDL

11 TROCRAOTPR

12 AOLIGNRI

14 NLRSOAITNTA

16 AGIME

18 TEERNGANMLE

19 EGARTINL

20 EERERTPIM

24 GURFIE

25 SSA

28 CSALE

29 CTIRELFEON

Down

1 OEVPR

2 EGLAN

3 RITOA

5 CNUENORGT

6 GHNMCTIA

7 LAERIRDTUQAAL

8 ETUNPSEOHY

13 TIONOATR

15 MPEIRPOSUES

17 ERECNT

21 HSR

22 IDSE

23 ETTS

26 SSS

27 ASA

13 CONGRUENT FIGURES

CONGRUENT IS A MATHEMATICAL WORD THAT MEANS 'IDENTICAL' AND '≅' IS THE SYMBOL FOR 'IS CONGRUENT TO'.

Name: _____

Due date: _____

Parent's signature: _____

Part A	/ 8 marks
Part B	/ 8 marks
Part C	/ 8 marks
Part D	/ 8 marks
Total	/ 32 marks

PART A: MENTAL MATHS

🚫 Calculators not allowed

1   Round 18 h 24 min to the nearest hour.

_____

2   Find 15% of $80. _____

3   True or false? The diagonals of a rectangle cross at right angles.

_____

4   Find x.

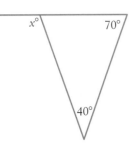

_____

5   Evaluate:

a $22 \div \dfrac{11}{13}$ _____

b $345.2 \div 1000$ _____

6   Simplify $\dfrac{20a^5b^2}{4ab^3}$. _____

7   Convert 0.2% to a decimal.

_____

PART B: REVIEW

1   (2 marks) Name 2 pairs of congruent triangles.

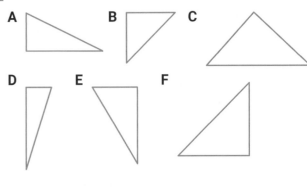

_____

2   These 2 trapeziums are congruent.

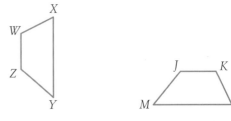

a   Which transformation has been performed on WXYZ to make JKLM?

_____

b   Which side in JKLM matches with WZ?

_____

c   Which angle matches with ∠Y? _____

d   Complete using the correct order of vertices:

WXYZ ≡ _____

9780170454537

3 True or false?

a Congruent figures have exactly the same size and shape. _____

b Matching sides and angles in congruent figures are equal. _____

PART C: PRACTICE

📝 › Congruent figures
› Tests for congruent triangles

1 For these congruent triangles, list:

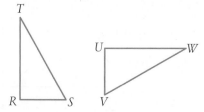

a all pairs of matching equal angles

_____

b all pairs of matching equal sides

_____

2 Which test proves that each pair of triangles is congruent?

a

_____

b

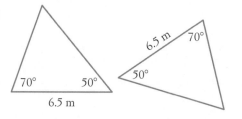

_____

3 The diagonals of a rhombus divide the rhombus into 4 congruent triangles.

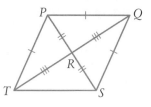

a Which test proves that the 4 triangles are congruent? _____

b Hence find the sizes of the 4 angles at point R. _____

c What does this prove about the diagonals of a rhombus? _____

d Are the diagonals of a rhombus equal?

_____

PART D: NUMERACY AND LITERACY

1 In the congruence test RHS, what is the meaning of:

a S? _____

b H? _____

2 a Which test proves that these 2 triangles are congruent? _____

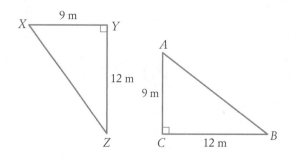

b Complete using the correct order of vertices:

$\triangle XYZ \equiv \triangle$ _____

3 (4 marks) The axis of symmetry DB of this kite divides it into 2 congruent triangles.

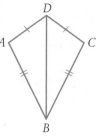

a List all pairs of matching equal angles.

_____

_____

b Complete: In a kite, one pair of opposite _____ are equal and one diagonal bisects 2 _____ of the kite.

(13) SIMILAR FIGURES 1

IN GEOMETRY, SIMILAR DOESN'T MEAN 'ALMOST THE SAME'. IT MEANS AN ENLARGEMENT OF REDUCTION. SAME SHAPE, BUT DIFFERENT SIZE.

Name:

Due date:

Parent's signature:

Part A	/ 8 marks
Part B	/ 8 marks
Part C	/ 8 marks
Part D	/ 8 marks
Total	/ 32 marks

PART A: MENTAL MATHS

🚫 Calculators not allowed

1 Evaluate $\dfrac{3 \times 4^2}{6}$. _____

2 What is 25% of one day in hours? _____

3 Simplify $8x - x^2 + 6x^2 - 2x$. _____

4 Find x.

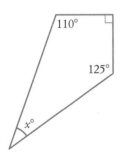

_____

5 What does the formula $V = \pi r^2 h$ describe?

_____

6 Write as a decimal the probability of 'even chance'. _____

7 Expand $(4p - 1)(p + 8)$.

_____

_____

8 Find the surface area of a cube with side length 5 cm. _____

PART B: REVIEW

1 Which test proves that these 2 triangles are congruent?

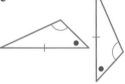

_____

2 For the SAS test, what type of angle is A?

_____

_____

3 (3 marks) Prove that $\triangle MNP \equiv \triangle ZXY$.

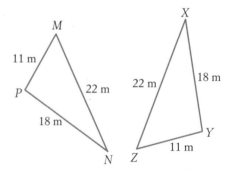

_____

_____

_____

_____

4 The diagonals of a rhombus divide the rhombus into 4 congruent triangles.

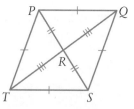

a Mark with a dot ∠QPR and the 3 angles that are equal to it.

b Mark with a cross ∠PQR and the 3 angles that are equal to it.

c What does this prove about the diagonals of a rhombus?

_____

_____

PART C: PRACTICE

📝 › Similar figures
    › Scale diagrams

1 Find the scale factor for each pair of similar figures.

a

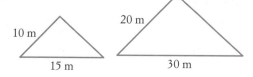

_____

b

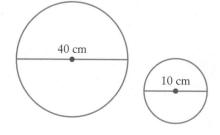

_____

2 Find in millimetres the actual length of each object.

a Scale 1 : 4

_____

b Scale 5 : 1

_____

3 For these similar triangles, find:

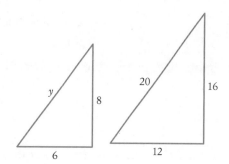

a the scale factor _____

b the value of y. _____

4 Ayon is 1.8 m tall and casts a shadow of 1.5 m. Find h, the height of a pole that casts a shadow of 7 m.

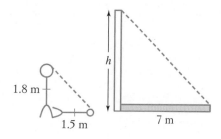

_____

_____

5 Find p for these similar triangles.

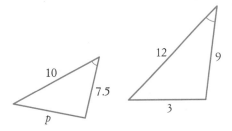

_____

_____

PART D: NUMERACY AND LITERACY

1 True or false?

a Similar figures have exactly the same size and shape.

_____

b Matching angles in similar figures are equal.

_____

c All circles are similar. _____

d All rhombuses are similar. _____

2 How does the size of an image compare to the original if its scale factor is between 0 and 1?

_____

3 **a** Which test proves that these 2 triangles are similar? _____

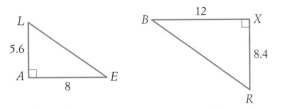

b What is the scale factor? _____

c Which angle matches with $\angle E$? _____

d Complete using the correct order of vertices:

$\triangle LEA \; ||| \; \triangle$ _____

9780170454537

Name:

Due date:

Parent's signature:

Part A	/ 8 marks
Part B	/ 8 marks
Part C	/ 8 marks
Part D	/ 8 marks
Total	/ 32 marks

SIMILAR FIGURES 2 (13)

THERE ARE 4 TESTS FOR PROVING
2 TRIANGLES ARE SIMILAR.
CAN YOU LIST THEM?

PART A: MENTAL MATHS

🚫 Calculators not allowed

1 Evaluate:

a $\dfrac{6}{7} - \dfrac{1}{3}$ _____

b 16×99 _____

2 Mark a pair of cointerior angles.

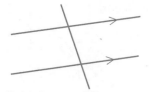

3 Expand and simplify $6(a - b) - 2(2a + b)$.

_____

4 In statistics, which measure of centre can have more than one value?

_____

5 Write 3 consecutive odd numbers beginning with x.

_____

6 What is the complementary event to a traffic light showing red?

_____

_____

7 Find 70% of $800. _____

8 Factorise $-7xy - 21y^2$. _____

PART B: REVIEW

1 For similar figures, write the property of their:

a matching angles

_____

b matching sides

_____

2 Draw the image of each shape using the scale factor.

a Scale factor = 4

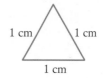

b Scale factor = $\dfrac{1}{2}$

3 Find the variable in each pair of similar figures.

a

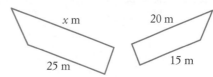

_____

b

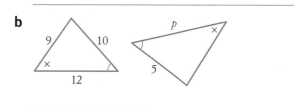

_____

HW HOMEWORK

4 These 2 trapeziums are similar.

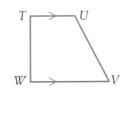

a Which angle in *TUVW* matches with ∠Q?

_____

b Which side matches with *PS*? _____

PART C: **PRACTICE**

 › Similar figures
› Tests for similar figures

1 In △*ABC* and △*ADE*, ∠A is common.

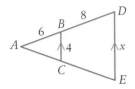

a Why is ∠*ACB* = ∠*AED*?

_____

b Which test can be used to prove that

△*ABC* ||| △*ADE*? _____

c Find the value of *x*.

_____

_____

2 a Why is ∠*LOM* = ∠*NOP*?

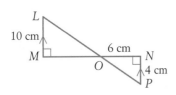

_____

b If the 2 triangles are similar, then complete:

△*LOM* ||| △ _____

c Find the length of *MO*.

_____

_____

d Find $\dfrac{\text{Area } \triangle NOP}{\text{Area } \triangle LOP}$.

_____

_____

3 Find *p*.

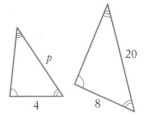

_____

PART D: **NUMERACY AND LITERACY**

1 What is the scale factor between congruent

figures? _____

2 What does '|||' mean in words?

_____

3 How far (correct to one decimal place) does the ladder reach up the wall?

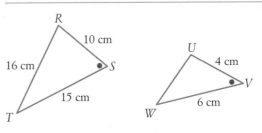

_____

4 List the 4 tests for congruent triangles.

_____

_____

5 What happens to a shape that is changed by a scale factor of $\dfrac{1}{3}$?

_____

6 a Which test proves that △*RST* ||| △*UVW*?

_____

b Find the scale factor. _____

c Find the length of *UW*. _____

9780170454537

Part A	/ 8 marks
Part B	/ 8 marks
Part C	/ 8 marks
Part D	/ 8 marks
Total	/ 32 marks

CONGRUENT (13) AND SIMILAR FIGURES REVISION

WE'RE UP TO THE END OF THIS TOPIC, AND THE QUESTIONS ARE GETTING QUITE COMPLEX. YOU MAY NEED TO ASK FOR HELP. KEEP AT IT!

PART A: MENTAL MATHS

 Calculators not allowed

1 Evaluate:

a $66\frac{2}{3}\%$ of $360 _____

b $-8 \times (-9)$ _____

2 Find a simplified algebraic expression for the perimeter of this triangle.

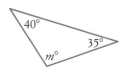

_____

3 Express $\frac{23}{4}$ as a mixed numeral.

_____

4 Find the average of -4, 6, 8 and -2. _____

5 What does the formula $A = \frac{1}{2}(a+b)h$ describe?

_____

6 Find m.

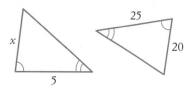

_____

7 Simplify $\dfrac{\left(x^2\right)^4}{x^3}$ _____

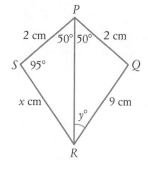

PART B: REVIEW

1 a (3 marks) Prove that $\triangle PSR \equiv \triangle PQR$.

_____
_____
_____
_____

b (2 marks) Find x and y.

_____

c What type of shape is $PQRS$? Why?

_____
_____

2 a Which test proves that these triangles are similar?

_____

b Find x.

_____

PART C: **PRACTICE**

📝 › Congruent and similar figures revision

1 Which test proves that each pair of triangles are congruent?

a

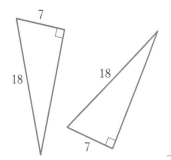

_____

b

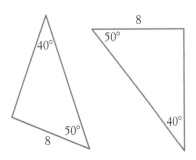

_____

2 In △ABE and △ACD, ∠A is common.

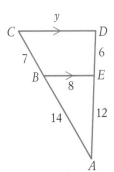

a Name a pair of equal angles.

_____

b Which test can be used to prove that
△ABE ||| △ACD? _____

c Find the value of y.

_____

_____

3 **a** Which test proves that these 2 triangles are congruent?

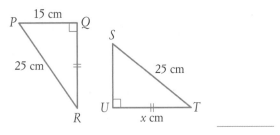

_____

b Complete: △PQR ≡ △ _____

c Find x.

_____

_____

PART D: **NUMERACY AND LITERACY**

1 In the diagram, state whether each pair of triangles given are congruent and, if so, state the test used.

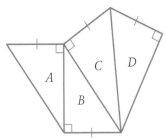

a A and B _____

b C and D _____

c A and C _____

9780170454537

2 a (2 marks) Explain why these 2 triangles are similar.

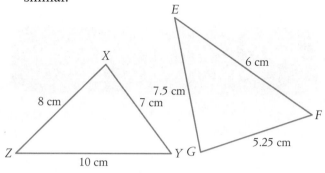

_____

_____

_____

b Complete: $\triangle XYZ \,|||\, \triangle$ _____

3 True or false?

a All equilateral triangles are similar.

_____

b All right-angled triangles are similar.

_____

SO LET'S GET GOING.

9780170454537

Chapter 1

StartUp assignment 1 PAGE 02

1 December **2** 2.9

3

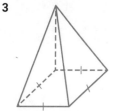

4 36

5 $11.25 **6** rhombus

7 1500 **8** $10

9 lines which are the same distance apart

10 4 : 3 **11** 25 cm

12 $0.3\dot{6}$ **13** 20

14 80% **15** 130

16 a $-4y$ **b** m^3

   c 12

17 a -1 **b** -30

18 a 32 **b** 7

   c 7

19 $\dfrac{1}{5}$

20 a 5 **b** 9

   c -9

21 10

22 a $3p + 6$ **b** $8k - 12$

23 $2n - 4$ **24** $9, 1, -5, -1, -7$

25 $y = 3x + 1$ **26** $9, 43, 201$

Challenge: Yes; 30, 55, 204

Algebra review PAGE 04

1 a $5u + 7$ **b** $2xy + 4b$

   c $2n^2$ **d** $9t - 5t^2$

   e $8m - 5$ **f** $3de + e$

   g $x + 2y$ **h** $4ab$

2 a $4d$ **b** $10 - k$

   c $7y$ **d** $t + 14$

   e $100 - 6m$ **f** $n + 1$

3 a $21de$ **b** $2p^2q^2$

   c a^3 **d** $16r^2$

   e kp^2r **f** $9b^4$

   g $12m^6$ **h** $6c$

   i $\dfrac{2}{y}$ **j** $\dfrac{3m}{2n}$

   k 10 **l** $\dfrac{a^2}{2}$

4 a 6 **b** 10

   c 20 **d** $-\dfrac{1}{2}$

   e -11 **f** -8

5 25 cm²

6 a $x^2 - 3x$ **b** $-8x - 24$

   c $uw - 4u$ **d** $-12 + 10b$

   e $6y + 18$ **f** $5 - 2r$

   g $a^2 + 8a - 6$ **h** q

7 a $6x - 14$ **b** $2x^2 - 14x$

8 62 km

9 a $\dfrac{x}{5}$ **b** $\dfrac{11u}{15}$

   c $\dfrac{7m}{8}$ **d** $\dfrac{a}{6}$

   e $\dfrac{5}{4d}$ **f** $\dfrac{wy}{5}$

   g $24p$ **h** $\dfrac{9r}{10}$

   i $\dfrac{8a}{3c}$

Algebra crossword PAGE 06

Across

1 sum **7** brackets

8 algebra **9** factor

10 fraction **13** decrease

18 quotient **19** denominator

21 increase **23** expression

25 symbols **26** like

27 rule **28** terms

29 factorise **30** expand

31 reciprocal

Down

1 square **2** highest

3 table **4** perimeter

5 variable **6** divide

11 consecutive **12** subtract

14 grouping **15** substitute

16 multiply **17** difference

20 pronumeral **22** evaluate

24 common

Substitution puzzle PAGE 14

1 H	**2** O	**3** R	**4** D	**5** W	**6** M
7 K	**8** U	**9** E	**10** A	**11** O	**12** L
13 X	**14** E	**15** Z	**16** P	**17** S	**18** Y
19 N	**20** T	**21** I	**22** G	**23** C	**24** V
25 R	**26** B	**27** N	**28** P	**29** F	**30** T

Answer: If a pronumeral is something that stands in place of a numeral, then is a protractor something that stands in place of a farming vehicle?

Chapter 2

StartUp assignment 2

1 $\frac{5}{7}$

2 3 for $6.95

3 a $5(3a - 7)$ **b** $2(6xy + 7x - 4y)$

4 10 000

5 $12\frac{1}{2}\%$

6 a $6x - 6$ **b** $2x(x - 3)$

7 12

8 Teacher to check

9 m^4

10 24

11 $m + 10mn$

12 9:5

13 $x = 104$

14 1, 4, 9, 16, 25, 36

15 270°

16 Teacher to check

17 Teacher to check

18 90°, 45°, 45°

19 a $x = 153$ **b** $x = 34$

20 a 400   **b** 7.29

    c 2.72   **d** 1.5

    e 26

21 a m **b** w

22 $r = 57$

23 a false   **b** true

    c false

24 a 4.90   **b** 12.08

    c 10.39   **d** 9.44

25 a true   **b** false

    c true

Challenge:

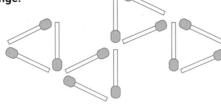

or

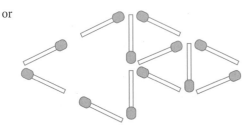

A page of right-angled triangles

1 $w = 8, x = 10, y = 6$ **2** $p = 4, r = 8.5, t = 7.5$

3 $a = 2.5, b = 2, c = 1.5$ **4** $d = 3$

5 $x = 4.5, y = 6, z = 7.5$ **6** $u = 7$

7 $m = 3$ **8** $k = 2$

9 $n = 5$ **10** $x = 1.5$

11 $t = 4.5$ **12** $s = 7.5$

13 $w = 3.5$ **14** $f = 5, n = 3, q = 4$

15 $y = 4$ **16** $d = 6.5, e = 6, f = 2.5$

Finding an unknown side

1 34   **2** 8.5

3 13   **4** 7

5 65   **6** 48

---

7 30.81   **8** 30.40

9 13.23   **10** 8.06

11 11.18   **12** 8.66

13 0.8   **14** 16.16

15 5.66   **16** 74.16

17 10.69   **18** 4.61

19 1.7   **20** 26.08

21 2.4

Chapter 3

StartUp assignment 3

1 7:5   **2** 360°

3 6.71

4 a 9 cm²   **b** 15.71 cm

5 Teacher to check   **6** $-3a - 24$

7 10   **8** 14

9 -8 **10** $-12m^2$

11 1, 2, 3, 6   **12** $\frac{2}{3}$

13 $\frac{19}{20}$ **14** $\frac{1}{6}$

15 a 0.75   **b** 0.$\dot{6}$

    c 0.09   **d** 0.1$\dot{6}$

16 a $\frac{2}{5}$ **b** $\frac{7}{20}$

17 a $\frac{2}{5}$ **b** $\frac{1}{25}$

    c $\frac{1}{8}$ **d** $\frac{7}{10}$

18 a 39   **b** 38.77

19 a 0.084   **b** 12.9

20 a 42   **b** $37\frac{1}{2}$

21 a 40%   **b** $33\frac{1}{3}\%$

    c 50%   **d** 3.75%

22 $8.50   **23** 24

24 $100 and $350   **25** $1.90

26 $47.83

Challenge: 81

Discounts and special offers

1 10-pack

2 a $140.25   **b** $44.20

    c $66.30   **d** $408

3 22.5%   **4** $38.70

5 J-Mart

6 a 14.1%   **b** 8%

    c 21.0%   **d** 16.8%

7 a $22.95   **b** 20%

8 a Large

    b Cheaper to produce in bulk

9 Hungry Pack   **10** 18.5%

11 a $414   **b** $560

    c $281   **d** $95

12 a Balmy World   **b** $921.82

13 a $632   **b** $2093.60

     c $197.60   **d** 10.4%

14 18.7%   **15** $83.25

16 Snappy

17 a $284   **b** 24.3%

18 8.9%   **19** $160.65

20 Family pack

Time calculations
PAGE 32

1 a 06:30   **b** 16:20

     c 23:05   **d** 14:45

     e 19:18   **f** 10:56

2 a 7:05 a.m.   **b** 6:55 p.m.

     c 12:30 p.m.   **d** 8:30 p.m.

     e 3:44 a.m.   **f** 3:17 p.m.

3 a 9:00 p.m.   **b** 1:00 p.m.

     c 07:00   **d** 3:00 p.m.

     e 09:00   **f** 8:00 a.m.

     g 10:30 p.m.   **h** 05:40

     i 2:30 a.m.

4 a 9 h 30 min   **b** 4 h 18 min

     c 12 h 36 min   **d** 7 h 24 min

     e 15 h 42 min   **f** 10 h 48 min

     g 5 h 15 min   **h** 8 h 18 min

5 a 6 h 18 min   **b** 3 h 56 min

     c 2 h 53 min   **d** 11 h 25 min

     e 7 h 7 min   **f** 6 h 44 min

6 a 4:35 a.m.   **b** 16:06

     c 10:14 p.m.   **d** 8:25 a.m.

7 a 8 h   **b** 8 h 40 min

     c 5 h 20 min   **d** 7 h 40 min

     e 5 h 13 min   **f** 10 h 43 min

8 a 3:40 a.m.   **b** 12:41 p.m.

     c 9:19 a.m.   **d** 10:06 a.m.

     e 01:06   **f** 11:21

9 a 3:30 p.m.   **b** 3:00 p.m.

     c 1:30 p.m.   **d** 3:30 p.m.

     e 3:30 p.m.   **f** 3:00 p.m.

10 a 4:00 a.m.   **b** 8:30 p.m.

     c 1:45 p.m.   **d** 10:47 a.m.

     e 15:35   **f** 07:20

Numbers crossword
PAGE 34

Down

1 third   **2** tenth

5 ascending   **7** rational

9 round   **10** denominator

13 twentyfive   **14** descending

15 evaluate   **17** increase

18 fraction   **22** decrease

26 places   **27** fifty

Across

1 ten   **3** decimal

4 numerator   **6** profit

8 hundred   **11** hundredths

---

12 quantity   **16** percentage

19 half   **20** cost

21 cent   **23** terminating

24 estimate   **25** simplify

28 recurring   **29** convert

30 discount   **31** selling

Chapter 4

StartUp assignment 4
PAGE 42

1 7 : 5   **2** $22 325

3 $330, $180   **4** $m^2 + 6m$

5 $\dfrac{1}{25}$ **6** 100 000

7 False   **8** 2 cents

9 10   **10** 16.5

11 126°   **12** 8

13 2100

14 a 9.38 cm   **b** 42.21 cm^2

15 360°   **16** 180°

17 a 19.6   **b** 81

     c 18   **d** 14.4

18 a 30°   **b** $\sqrt{3}$

19 48°   **20** 18°

21 a $x = 35$ **b** $h = 20$

     c $d = 5$ **d** $y = 2.5$

     e $p = 2.25$

22 w **23** $90° - \alpha°$

24 19 min

25 a $x = 153$ **b** $x = 34$

26 a $n = 12$ **b** $\angle T$

27 7 h 9 min

28 a $d = 1$ **b** $h = 1.41$

Challenge:

One solution is: Balance 6 coins and 6 coins to find which side holds the counterfeit coin. Then balance 3 coins and 3 coins from the side holding the counterfeit to find which group of three coins includes the counterfeit. Then choose two coins from the coins containing the counterfeit and balance 1 and 1. If they balance, then the third coin is the counterfeit. If they don't, then the lighter one is the counterfeit.

Trigonometric calculations
PAGE 44

1 a 65°   **b** 46°

     c 18°   **d** 29°

     e 73°   **f** 60°

2 a 18°34′   **b** 79°7′

     c 40°14′   **d** 36°48′

     e 61°00′   **f** 8°15′

3 a 17°30′   **b** 3°24′

     c 46°8′   **d** 32°42′

     e 70°13′   **f** 81°54′

     g 73°49′   **h** 22°24′

     i 49°22′   **j** 56°25′

4 a 8.25°   **b** 26.20°

     c 37.62°   **d** 13.35°

     e 40.30°   **f** 77.07°

5 a 0.435 **b** 0.293
  c 0.982 **d** 1.698
  e 0.5 **f** 6.889
  g 7.173 **h** 7.065
  i 15.980 **j** 0.448
  k 26.870 **l** 6.592
  m 17.194 **n** 12.926
  o 3 **p** 8.327
6 a false **b** true
  c false **d** true
  e true **f** true
  g false **h** true
  i false **j** true
7 a 25° **b** 36°
  c 12° **d** 39°
  e 35° **f** 13°
8 a 53.13° **b** 69.82°
  c 58.36° **d** 8.63°
  e 51.32° **f** 59.74°
9 a 72°54′ **b** 41°59′
  c 46°53′ **d** 85°36′
  e 36°52′ **f** 84°16′

Finding an unknown side (2) PAGE 46

1 7.73 m **2** 29.14 km
3 2.23 cm **4** 18.89 m
5 17.27 cm **6** 176.79 m
7 4.50 cm **8** 12.73 m
9 5.16 km **10** 14.86 m
11 15.07 cm **12** 92.04 km
13 31.98 m **14** 16.68 cm
15 31.70 m **16** 27.07 km
17 11.66 m **18** 19.37 mm
19 290.84 km **20** 45.83 m
21 96.77 m

Finding an unknown angle PAGE 48

1 34° **2** 61°
3 39° **4** 48°
5 54° **6** 41°
7 59° **8** 45°
9 57° **10** 54°
11 61° **12** 39°
13 36° **14** 63°
15 37° **16** 73°
17 27° **18** 50°
19 37° **20** 36°
21 11°

Trigonometry crossword PAGE 50

Across

5 adjacent **7** tangent
9 cosine **12** sine
14 inverse **16** right angled
18 hypotenuse **19** angle
20 opposite

Down

1 perpendicular **2** theta
3 vertex **4** minute
6 triangle **8** similar
9 cos **10** diagonal
11 degree **12** side
13 sixty **15** distance
17 ratios

Chapter 5

StartUp assignment 5 PAGE 60

1 $84 **2** 135
3 19 **4** $3x$
5 162 **6** 7.7
7 105 km **8** 6
9 3.464 **10** parallelogram
11 3.071 **12** 4
13 2.25 cm **14** 50
15 $a^2 + 9a + 20$
16 a 243 **b** 4913
  c 625 **d** -8
  e 3400 **f** 0.0034
  g 24 **h** 25.5
  i 8 **j** 6
17 a 10^5 **b** 3^8
  c 2^5 **d** 3^6
18 a b^2 **b** m^4
  c n^3 **d** $6x^4$
  e a **f** 5
  g u^4 **h** $4f^8$
19 a 37 500 **b** 9200
  c 0.08

Challenge: In rows:

3	2	4
2	H	2
4	2	3

Other solutions are possible.

Index laws review PAGE 62

1 a u^8 **b** x^5
  c b^6 **d** t^7
  e $24r^6$ **f** $18m^3$
  g $-10a^8$ **h** $-4d^{13}$
  i $-16e^4$ **j** $36d^6$
2 a f^2 **b** k^3
  c q^8 **d** v^5
  e $10p$ **f** $-6r^5$
  g $8e^3$ **h** $-3u^3y$

i $-4p^5z^3$

j $\dfrac{9}{a^2}$

k $\dfrac{4k}{5h}$

l $\dfrac{-2}{3d^2}$

3 a r^{20}

b d^6

c u^9

d a^{14}

e $\dfrac{1}{t^2}$

f y^2

g $8u^9$

h $27r^6$

i $\dfrac{1}{5u^5}$

j m^8

k $4w^{12}$

l $\dfrac{1}{16d^4}$

4 a 1

b 1

c 1

d 1

e 3

f 2

g 1

h 1

i 1

j a

k -7

l 1

m $2d$

n $20m$

o 1

p -1

5 a $\dfrac{1}{r^2}$

b $\dfrac{1}{v}$

c $\dfrac{1}{m^3}$

d $\dfrac{1}{p^4}$

e $\dfrac{3}{u}$

f $\dfrac{10}{t^4}$

g $\dfrac{8}{a^2}$

h $\dfrac{5}{d^3}$

i $\dfrac{1}{ab}$

j $\dfrac{1}{4k^2}$

k $\dfrac{4k}{r}$

l $\dfrac{3}{x^2 y}$

m $-\dfrac{1}{n}$

n $\dfrac{1}{q^2}$

o $-\dfrac{2}{b}$

p $-\dfrac{1}{d^3}$

6 a $a^3 b^3$

b $f^{20} g^8$

c $-8e^6 p^9$

d $-k^5 m^{20} n^{25}$

e $\dfrac{a^3}{8}$

f $\dfrac{b^4}{25}$

g $\dfrac{r^{20}}{p^{12}}$

h $\dfrac{243a^{30}}{b^5 c^{10}}$

7 a u^{-2}

b 5^{-3}

c 3^{11}

d 2^{13}

e 5^8

f $-7f^{-2}$

g $10a^{-4}$

h 3^9

i 4^2

j $-a^{-1}$

k $6a^{-1}$

l $5(ab)^{-1}$

8 a $\dfrac{1}{6}$

b $\dfrac{1}{25}$

c 279 936

d $\dfrac{9}{25}$

e 1

f 59 049

g $\dfrac{-64}{343}$

h 1

Scientific notation puzzle PAGE 64

1 W	**2** O	**3** A	**4** I	**5** T	**6** P
7 C	**8** W	**9** Z	**10** K	**11** T	**12** E
13 N	**14** S	**15** L	**16** H	**18** B	**19** U
20 O	**21** S	**22** V	**23** F	**24** X	**25** U
26 E	**27** Y	**28** N	**29** G	**30** R	

When you work with scientific notation, you get to use the powers of ten!

Indices crossword PAGE 66

Across

- **2** power
- **5** one
- **6** negative
- **7** zero
- **8** scientific
- **9** estimate
- **11** law
- **14** form
- **15** reciprocal
- **20** base
- **21** significant
- **22** add
- **24** index
- **25** figures
- **27** ascending
- **29** exponent

Down

- **1** integer
- **2** product
- **3** quotient
- **4** indices
- **10** multiply
- **12** notation
- **13** descending
- **16** root
- **17** cubed
- **18** divide
- **19** ten
- **23** subtract
- **26** expanded
- **28** squared

Chapter 6

StartUp assignment PAGE 76

1 9 months

2 -64

3 12.85 cm

4 4

5 a 17 cm

b 60 cm^2

6 3.162

7 $6m^2$

8 Dean $275, Jerry $165

9 $x = 4$

10 5000

11 triangular prism

12 -8

13 $47.84

14 $8m - 20$

15 30

16 a

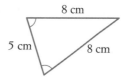

b $\angle DEC$

17 50

18 $a = 70, b = 70, c = 110$

19 An angle that measures between 90° and 180°

20 a alternate

b corresponding

21 2

22 $w = 65$

23 $e = 105$

24 60°

25 $m = 95$

26

5 cm 8 cm 8 cm

WORKSHEET AND PUZZLE SHEET ANSWERS

27

28

29 130°

30 No

31 $t = 70$

32

33 $u = 110$

34

Challenge:

Naming quadrilaterals

PAGE 78

1 Teacher to check diagrams

2 a true **b** true
  c false **d** true
  e true

3 a true **b** false
  c false **d** false
  e false

4 a false **b** false
  c true **d** false
  e false

5 a false **b** true
  c false **d** true
  e false

6 a parallelogram, rhombus, rectangle, square
  b parallelogram, rhombus, rectangle, square

7 a rectangle, square **b** kite, rhombus, square
  c rectangle, square **d** rhombus, square, kite
  e parallelogram, rhombus, rectangle, square
  f rhombus, square **g** rhombus, square
  h rhombus, square **i** rhombus, rectangle
  j parallelogram, rhombus, rectangle, square

8 a rhombus **b** rectangle
  c parallelogram **d** rhombus
  e kite **f** rectangle
  g parallelogram **h** parallelogram
  i trapezium **j** rectangle

Find the unknown angle

PAGE 80

1 $a = 22$ **2** $b = 81$
3 $c = 138$ **4** $d = 111$
5 $e = 295$ **6** $f = 108$
7 $g = 29$ **8** $h = 77$
9 $i = 74$ **10** $j = 25$
11 $k = 64$ **12** $l = 45$
13 $m = 99$ **14** $n = 60$
15 $o = 15$ **16** $p = 136$
17 $q = 124$ **18** $r = 127$
19 $s = 65$ **20** $t = 37$

Geometry crossword

PAGE 82

Across
  4 vertically **7** cointerior
  10 angles **12** convex
  14 complementary **18** regular
  19 scalene **21** trapezium
  23 square **25** equilateral
  28 perpendicular

Down
  1 isosceles **2** parallelogram
  3 axis **5** bisect
  6 triangle **8** equal
  9 kite **11** sixty
  13 quadrilateral **15** one
  16 parallel **17** none
  20 exterior **22** diagonal
  24 rectangle **26** adjacent
  27 decagon **29** polygon

Chapter 7

StartUp assignment 7

PAGE 88

1 $22.75 **2** $20p + 4$
3 $10g^4$ **4** 0.1202
5 a 11 **b** 53°
6 $\frac{9}{40}$ **7** 5
8 0.8$\dot{3}$
9 One of the following: bisect each other, cross at right angles, they are axes of symmetry
10 $y = 30$ **11** $213.15
12 $3x^2 + 11x - 20$ **13** 288 cm²
14 385 cm³
15 a Yes **b** Yes
16 a No **b** No
17 Teacher to check
18 a 40 **b** 20
  c 8
19 a $14k - 7$ **b** $-9k + 15$
  c $2x$ **d** $15d$
  e $8 - 6t$
20 a $h - 4$ **b** $2x + 1$
  c $p - 2$

21 a $x = -4$ **b** $p = 13$
    c $k = 36$ **d** $m = 4\frac{1}{2}$
    e $a = 4$ **f** $y = 2$

Challenge: 13

Word problems with equations PAGE 90

1 31   **2** 12
3 32   **4** 6 cm, 11 cm, 13 cm
5 49 and 50   **6** 20
7 Harry $31, Nicky $89   **8** 24
9 14 cm $\times$ 18 cm   **10** 32, 33, 34
11 13 $\times$ 10c, 15 $\times$ 20c   **12** octagon (eight-sided)
13 15   **14** 10
15 98°   **16** 19
17 30, 32, 34   **18** 36
19 8   **20** 75c
21 55°   **22** 5
23 36   **24** 32 chickens, 9 pigs
25 Mrs Grant 56, Tess 36, Carly 34 and Troy 30

Working with formulas PAGE 92

1 40.375 m
2 a 112.5   **b** 14
3 4.23×10^{-3} **4** 62 km
5 a 1.83 m   **b** 32.81 ft
6 a 131 matches   **b** 27 triangles
7 a $m = 3s + 1$ **b** 46 matchsticks
8 7.5 cm   **9** 96
10 a 1190 m^3 **b** 60.7 cm
11 a 15.625 miles   **b** 12.8 km
12 7.00 cm   **13** 12 m
14 48
15 a 100°C   **b** 392°F
16 3.24 seconds   **17** 10 sides
18 a $266.40   **b** $c = 185 + 3.7d$

Equations crossword PAGE 100

Across

2 variable   **3** undoing
5 check   **7** linear
10 consecutive   **12** fraction
13 test   **16** equation
17 multiple   **18** square
20 LHS   **21** cubic
22 quadratic   **23** RHS
24 inverse   **25** surd
26 equal

Down

1 pronumeral   **4** number
6 unknown   **8** root
9 LCM   **11** subject
14 solve   **15** operation
18 substitute   **19** algebra

Chapter 8

StartUp assignment 8 PAGE 101

1 $a^2 - 4a$ **2** 16.5
3 12 cm^2 **4** $4x + y$
5 -5 **6** 3.142
7 $a^2 = b^2 + c^2$ **8** -2
9 $x = 110$ **10** $\frac{3}{5}$
11 $\frac{5}{8}$ **12** x-axis
13   **14** 5

15 alternate

16 a $\frac{1}{10}$ **b** $\frac{1}{3}$
17 a 0.48   **b** 0.2
18 $57.60   **19** 60%
20 $12.78
21 a $4937   **b** $1185
    c $162.90
22 365   **23** 26
24 168   **25** 92.5%
26 37.5%   **27** $1.16
28 $484.10   **29** 25.5%
30 $780   **31** $41\frac{2}{3}$%
32 $2664   **33** $70.20
34 8%   **35** $15 194
36 500

Challenge: Yes, on the 7th day

Earning money crossword PAGE 103

Across

3 piecework   **5** deduction
7 overtime   **13** double
14 allowable   **17** half
18 per annum   **19** gross
20 loading   **22** percentage
25 net   **27** PAYG

Down

1 fifty two   **2** month
4 retainer   **6** leave
8 taxable   **9** financial
10 salary   **11** fortnight
12 wage   **15** commission
16 income   **17** holiday
21 week   **23** earnings
24 annual   **26** tax

Time and money calculations PAGE 104

1 a $14.57   **b** $2.91
    c $50.40   **d** $244.08

2 a 0.14　　　　**b** 0.55
　c 0.08　　　　**d** 0.8

3 $40\frac{1}{2}$ hours

4 a $765.24　　　**b** $80.85
　c $626.68　　　**d** $500.72
　e $97.56　　　**f** $759.62
　g $467.20　　　**h** $635.60
　i $421.89

5 a $143.25　　　**b** $21.85
　c $44.30

6 $15 258

7 a 12　　　　**b** 4.5
　c 6　　　　**d** 10.5

8 a 12　　　　**b** 7
　c 14　　　　**d** 5
　e 26　　　　**f** 2

9 a 24%　　　　**b** 39.5%
　c 60%

10 $1.44

11 a $799.70　　　**b** $4114.80
　c $499.55　　　**d** $214.11

12 a $8726　　　**b** $644
　c $16 299

13 $582

14 a $168　　　**b** $4
　c $1456

15 a 41.5 hours　　**b** $22.04

16 $2091

17 a 8 hours 30 minutes　**b** 4 hours 30 minutes
　c 9 hours　　　　**d** 8 hours 30 minutes

18 $69.40

19 a $41.40　　　**b** 11.5%

20 $742

Percentages without calculators　　PAGE 106

1 a $\frac{7}{10}$　　　　**b** $\frac{3}{25}$
　c $\frac{2}{5}$　　　　**d** $\frac{1}{3}$
　e $\frac{1}{20}$　　　　**f** $\frac{3}{4}$

2 a 0.18　　　　**b** 0.09
　c 0.65　　　　**d** 0.2
　e 0.125　　　　**f** 0.883

3 a 25%　　　　**b** 20%
　c $66\frac{2}{3}$ %　　　**d** 90%
　e $12\frac{1}{2}$ %　　　**f** 60%

4 a $65　　　　**b** $11
　c $48　　　　**d** $4.50
　e $9　　　　**f** $149.60
　g $2.55　　　　**h** $140

5 a $336　　　　**b** $54.60
　c $200　　　　**d** $756

6 A

7 a 20%　　　　**b** 75%
　c $33\frac{1}{3}$ %　　　**d** 12%
　e 12.5%　　　　**f** 80%

8 a 3.2 L　　　　**b** $3.60
　c 40 min　　　**d** $7.50
　e 0.6　　　　**f** 3.75 kg
　g $3　　　　**h** $385

9 B

10 a $142.50　　　**b** $69.30
　c $50　　　　**d** $385

11 64　　　　**12** D

13 $300

14 a 60%　　　　**b** 25%
　c $12\frac{1}{2}$ %　　　**d** 20%
　e $12\frac{1}{2}$ %　　　**f** 80%

15 B　　　　**16** $300

17 $74.80　　　**18** $62\frac{1}{2}$ %

19 60　　　　**20** $63

21 $1380

22 a $32　　　　**b** 4%

23 900　　　　**24** 30%

25 15%

Wages and salaries　　PAGE 108

1 a $191.20　　　**b** $358.50
　c $908.20

2 $72.50　　　　**3** $752.40

4 $1032.75　　　**5** $2109.87

6 a $2287.12　　　**b** $9945.17
　c $4574.24

7 $821.60　　　**8** $107.25

9 T. Bagg: 35; $903.00
　B. Haiv: 31; $716.12
　K. Kuzmi: 31; $837.00
　P. Nutbutta: 31.5; $681.03
　C. Shore: 26.5; $898.35

10 a 18 hours　　　**b** $680.40

11 $57.00

12 a $91 043.68　　　**b** $1750.84
　c $7586.97

13 a $196　　　**b** 66 hours

14 $812.68

15 a $85 620　　　**b** $3281.72

16 a $2146.42　　　**b** $1466.99
　c $1154

17 a $67\frac{1}{2}$ hours　　**b** $7\frac{1}{2}$ hours

18 D　　　　**19** Jack by $7076

20 the fortnightly wage

Chapter 9

StartUp assignment 9

PAGE 112

1 -8　　**2** $\sqrt{7}$

3 $3b^2$　　**4** $360°$

5 $x = -3\frac{1}{2}$　　**6** 117.81

7 64 cm^3　　**8** $\frac{2}{3}$

9 $16y(1 - 4x)$　　**10** 2.5×10^{-4}

11 $-8x + 20$　　**12** 4900

13 22.5 min　　**14** $7 : 6$

15 $41°14'$

16 a sector graph (or pie chart)　**b** dot plot

　c column graph　**d** line graph

17 a 5.5　　**b** 7

18 a 4 and 6　　**b** 4

19 a 2.35　　**b** 2

20 10

21 a Frequency = 2, 5, 5, 6, 5, 1　**b** 24

　c 9　　**d** mode

　e 8.5　　**f** 6

　g 12　　**h** 5

　i

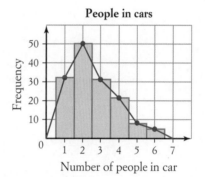

22 a $f = 1, 1, 0, 3, 3, 4, 2, 3, 3, 20; fx = 2, 3, 0, 15, 18, 28, 16,$
27, 30, 139

　b 6.95

Challenge: From top to bottom: Davis, Badger, Cooper, Aspinall, Ellsmore

Statistics review

PAGE 114

1 a discrete　　**b** continuous

　c discrete　　**d** continuous

2 a $4\frac{1}{4}$　　**b** 5

　c 5　　**d** 4

　e $\frac{3}{10}$

3 a 20　　**b** 6.4

　c 7　　**d** 7

　e 6　　**f** 65%

4 a 24　　**b** 19, 22, 24

　c 24　　**d** 26.1

　e 48　　**f** $\frac{1}{2}$

5 a

Score, x	Frequency, f	fx
1	4	4
2	6	12
3	7	21
4	8	32
5	5	25
6	6	36
Total		130

b 36　　**c** 3.61

d 4　　**e** 4

f $\frac{5}{18}$　　**g** 44.4%

6 a

Score, x	Frequency, f	Cumulative frequency
1	33	33
2	50	83
3	32	115
4	22	137
5	8	145
6	5	150

b 150　　**c** 2.58

d 2　　**e** 5

f 2　　**g** and **h**

People in cars

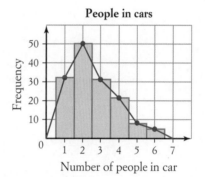

Number of people in car

Statistics crossword

PAGE 116

Across

1 symmetrical　　**3** histogram

6 leaf　　**9** positively

12 population　　**15** statistics

20 median　　**21** categorical

22 location　　**24** survey

25 census　　**26** skewed

27 mean　　**28** random

Down

2 cumulative　　**4** sample

5 bias　　**7** frequency

8 data　　**10** left

11 continuous　　**13** polygon

14 discrete　　**16** tail

17 plot　　**18** negatively

19 outlier　　**21** cluster

23 mode　　**26** stem

Mean, median and mode

PAGE 124

Set	Mean	Median	Mode(s)	Range
1	14.71	15	11	13
2	18.88	18	10 & 16	18
3	11.18	11	14	17

Set	Mean	Median	Mode(s)	Range
4	4.67	4	1	12
5	25.4	22.5	20	22
6	31.1	30	40	48
7	62.89	64	64	16
8	3.14	3	2 & 5	4
9	87	87	no mode	8
10	30	30	no mode	40
11	4.46	4	3 & 4	7
12	17	17	17	6
13	7	7	7	6
14	30.86	20	17	83
15	9.18	9	8	5
16	22.38	22.5	21 & 23	5
17	49.57	50	48 & 51	5
18	9.17	9	9	1
19	2	2	1	5
20	5.62	6	6	7

Chapter 10

StartUp assignment 10 PAGE 125

1 $(3, -1)$

2 5

3 $x^2 - 10x + 25$

4 1, 3, 5, 15

5 64

6 8

7 $x = 12.2$

8 co-interior

9 $x = 11$

10 25

11 12%

12 $6p^6$

13 130

14 $\dfrac{19}{20}$

15 25 L

16 1000

17 3600

18 a 20 cm **b** 21 cm²

19 14 cm²

20 7 hours 10 minutes

21 2

22 16 cm²

23 3.5 cm

24 a 18.85 cm **b** 28.27 cm²

25 a 18 cm **b** 16 cm²

26 a 7.5 **b** 18 cm

   c 13.5 cm²

27 a rectangular prism **b** 15 cm³

   c 6

28 $\dfrac{5}{12}$ **29** 5:30 a.m.

30 a triangular prism **b** 5

   c two right-angled triangles and three rectangles

31 trapezium

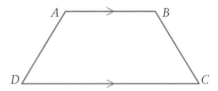

Challenge: 26 posts

Area ID PAGE 127

2 $A = 36$ cm²

3 $A = 10$ cm²

4 Parallelogram, $A = bh = 48$ cm²

5 Triangle, $A = \dfrac{1}{2} bh = 25$ cm²

6 Square, $A = s^2 = 36$ cm²

7 Circle, $A = \pi r^2 \approx 28.27$ cm²

8 Trapezium, $A = \dfrac{1}{2} (a + b)h = 68$ cm²

9 Rhombus, $A = \dfrac{1}{2} xy = 6$ cm²

10 parallelogram, $A = bh = 15$ cm²

11 Rectangle, $A = lw = 28$ cm²

12 Triangle, $A = \dfrac{1}{2} bh = 27$ cm²

13 Kite, $A = \dfrac{1}{2} xy = 16$ cm²

14 Circle, $A = \pi r^2 \approx 78.54$ cm²

15 Square, $A = s^2 = 64$ cm²

16 Trapezium, $A = \dfrac{1}{2} (a + b)h = 27.5$ cm²

17 Triangle, $A = \dfrac{1}{2} bh = 10.5$ cm²

18 Rhombus, $A = \dfrac{1}{2} xy = 22.5$ cm²

19 Circle, $A = \pi r^2 \approx 50.27$ cm²

20 Parallelogram, $A = bh = 18$ cm²

21 Trapezium, $A = \dfrac{1}{2} (a + b)h = 34$ cm²

22 Semi-circle, $A = \dfrac{1}{2} \pi r^2 \approx 76.97$ cm²

23 Kite, $A = \dfrac{1}{2} xy = 22$ cm²

24 Sector, $A = \dfrac{1}{3} \pi r^2 = 16.76$ cm²

A page of prisms and cylinders PAGE 128

(surface area, volume)

1 2950 cm²		8750 cm³
2 1782 mm²		3645 mm³
3 1.5 m²		0.125 m³
4 408 cm²		360 cm³
5 70.7 m²		42.4 m³
6 111.9 cm²		56 cm³

WORKSHEET AND PUZZLE SHEET ANSWERS   PS   WS

7 6283.2 mm²   37 699.1 mm³

8 1350 mm²   3375 mm³

9 51 cm²   15 cm³

10 3240 mm²   8100 mm³

11 879.6 cm²   1806.4 cm³

12 311.04 m²   373.248 m³

Surface area and volume crossword

Across

2 volume

7 base

12 dimensions

16 capacity

20 error

22 circle

26 triangle

30 trapezium

33 diagonal

4 radius

8 prism

13 limits

17 rhombus

21 kite

23 perimeter

28 accuracy

32 century

34 estimate

Down

1 hour

5 diameter

9 sector

11 composite

14 metric

18 hectare

24 rectangle

27 litre

31 mass

3 faces

6 square

10 circumference

13 length

15 surface

19 tonne

25 breadth

29 area

WORKSHEET AND PUZZLE SHEET ANSWERS

Chapter 11

StartUp assignment 11

1 **a** 612 000 000   **b** 6.12×10^8

2 $-6x - 10$

3 True

4 360°

5 $\dfrac{3}{20}$

6 $x = -4$

7 62 cm²

8 -1

9 $0.58\dot{3}$

10 $110.50

11 27.5 km

12 none

13 $1270.80

14 $115; $69

15 **a** $-\dfrac{1}{3}$ **b** -6

    c 10   **d** 5

16 15

17 $\dfrac{a+b}{2}$

18 **a** $(3, -2)$ **b** 3 units

19 **a** $\sqrt{34}$ **b** 5.83

20 2

21 **a** $-6, -2, 2, 6, 10$ **b** 12, 10, 8, 6, 4

22 $y = 3x + 3$

23 **a** $(2, 5)$ **b** 4 units

    c $\sqrt{20}$ units

Challenge: Fill the 9 L bucket, pour it into the 4 L bucket twice. The 1 L that's left, pour it into the 4 L bucket. Fill the 9 L again, pour it into the 4 L bucket but, now, only 3 L will fit so there will be 6 L left in the 9 L bucket.

A page of intervals

	Length	Midpoint	Gradient	y-intercept	Equation
1	5.66	$(-2, 2)$	1	4	$y = x + 4$
2	8.94	$(-4, 2)$	$\dfrac{1}{2}$	4	$y = \dfrac{1}{2}x + 4$
3	8	$(-8, 4)$	–	–	$x = -8$
4	2.83	$(-7, 7)$	-1	0	$y = -x$
5	6	$(-3, 6)$	0	6	$y = 6$
6	8.49	$(3, 3)$	-1	6	$y = -x + 6$
7	6.32	$(3, -1)$	$\dfrac{1}{3}$	-2	$y = \dfrac{1}{3}x - 2$
8	3	$\left(-1\dfrac{1}{2}, -2\right)$	0	-2	$y = -2$
9	1	$\left(-3, -2\dfrac{1}{2}\right)$	–	–	$x = -3$
10	3.61	$\left(-4\dfrac{1}{2}, -2\right)$	$-\dfrac{2}{3}$	-5	$y = -\dfrac{2}{3}x - 5$
11	4	$(-8, -1)$	0	0	$y = -1$
12	6.40	$\left(-7\dfrac{1}{2}, -3\right)$	$-\dfrac{4}{5}$	-9	$y = -\dfrac{4}{5}x - 9$
13	5.10	$\left(-5\dfrac{1}{2}, -7\dfrac{1}{2}\right)$	5	20	$y = 5x + 20$
14	6.08	$\left(-3, -9\dfrac{1}{2}\right)$	$\dfrac{1}{6}$	-9	$y = \dfrac{1}{6}x - 9$

9780170454537   **Answers**   197

	Length	Midpoint	Gradient	y-intercept	Equation
15	3.16	$(\frac{1}{2}, -7\frac{1}{2})$	3	-9	$y = 3x - 9$
16	4.12	$(1\frac{1}{2}, -8)$	-4	-2	$y = -4x - 2$
17	2.24	$(3, -9\frac{1}{2})$	$\frac{1}{2}$	-11	$y = \frac{1}{2}x - 11$
18	6.40	$(6, -6\frac{1}{2})$	$\frac{5}{4}$	-14	$y = \frac{5}{4}x - 14$
19	4.47	$(9, -2)$	2	-20	$y = 2x - 20$
20	6.32	$(9, 3)$	-3	30	$y = -3x + 30$
21	5	$(6, 4\frac{1}{2})$	$\frac{3}{4}$	0	$y = \frac{3}{4}x$
22	4	$(4, 5)$	–	–	$x = 4$
23	8.25	$(0, 8)$	$-\frac{1}{4}$	8	$y = -\frac{1}{4}x + 8$
24	14	$(3, 9)$	0	0	$y = 9$
25	5.39	$(7\frac{1}{2}, 8)$	$\frac{2}{5}$	5	$y = \frac{2}{5}x + 5$

Coordinate geometry crossword PAGE 146

Across

3 parabola
8 circle
10 constant
12 run
14 quadratic
19 intercept
23 symmetry
26 distance

4 graph
9 axes
11 curve
13 equation
16 linear
20 axis
24 line

Down

1 gradient
3 proportion
6 hyperbola
11 concave
17 midpoint
21 steepness
25 vertical

2 horizontal
5 rise
7 interval
15 radius
18 vertex
22 surd

Chapter 12

StartUp assignment 12 PAGE 156

1 $21 387
2 The most common data value(s)
3 1 : 5000
4 $b + 3$
5 75°
6 $n = 3$
7 $x = 44$
8 a 20.47 cm   b 26.18 cm²
9 $x = 3\frac{1}{2}$
10 horizontal
11 17 units
12 $290.23
13 a $\frac{1}{2}$ b 3

14 64%
15 a $\frac{2}{5}$ b $\frac{1}{5}$
   c $\frac{3}{10}$
16 a $\frac{1}{6}$ b $\frac{2}{3}$
   c $\frac{5}{6}$
17 $\frac{1}{4}$ 18 Teacher to check
19 a 0.55   b $\frac{1}{6}$
20 certain, must happen   21 0.5
22 Reds win, Blues win, draw   23 $\frac{2}{5}$
24 0.02   25 $\frac{5}{8}$
26 55%
27 a $\frac{1}{4}$ b $\frac{1}{13}$
   c $\frac{3}{13}$ d $\frac{5}{13}$
28 19.2%   **Challenge:** 24

Matching probabilities PAGE 158

1 P	2 A	3 I	4 Z	5 M	6 N
7 O	8 T	9 R	10 Y	11 E	12 K
13 D	14 J	15 U	16 F	17 L	18 A
19 G	20 H	21 Q	22 S	23 U	24 N
25 V	26 X	27 B	28 W	29 E	30 C

No chance at all. It's an Australian slang term, a pun on the name of an old Melbourne firm 'Buckleys and Nunn' (none).

Tree diagrams

PAGE 160

1 a 2   **b** $\frac{1}{2}$

2 a 4   **b** $\frac{1}{4}$

3 8   **4** $\frac{1}{8}$

5 a $\frac{3}{8}$ **b** $\frac{1}{8}$

  c $\frac{3}{8}$

6 16   **7** $\frac{1}{16}$

8 a $\frac{3}{8}$ **b** $\frac{1}{4}$

  c $\frac{1}{4}$

9 HHHH, HHHT, HHTH, HHTT, HTHH, HTHT, HTTH, HTTT, THHH, THHT, THTH, THTT, TTHH, TTHT, TTTH, TTTT

Probability crossword

PAGE 168

Across

  2 impossible

  7 space

12 theoretical

17 outcome

24 exclusive

27 dice

29 least

31 replacement

  6 two way

  8 frequency

14 most

22 trial

26 event

28 list

30 table

Down

  1 chance

  4 complementary

  9 expected

11 favourable

15 sample

18 two step

20 coin

23 random

30 tree

  3 mutually

  5 certain

10 die

13 experiment

16 probability

19 relative

21 Venn

25 diagram

Chapter 13

StartUp assignment 13

PAGE 169

1 $\frac{5}{24}$ **2** 2.6×10^{-2}

3 $1475.19   **4** 120°

5 $\frac{3}{W^2}$ **6** -5

7 $7 - m$ **8** 8.49

9 $8\frac{1}{2}$ **10** $k = 7$

11 $180° - 2x°$ **12** 30.8 cm

13 a 5   **b** 60 cm²

14 1.29 m

15 a 1 : 500   **b** 1 : 25

  c 1 : 40 000

16 a $x = 3$ **b** $h = 2$

  c $t = 6\frac{1}{2}$

17 a 15 km   **b** 10 cm

18 1 : 45   **19** 1 : 50 000

20 a 60   **b** 75

  c 38   **d** 126

  e 58

21 a reflection   **b** rotation

22 SSS, SAS, RHS, AAS

23 a 5   **b** $\angle W$

24 a Teacher to check   **b** 2.9 cm

Challenge: $W = 38, X = 42, Y = 20, Z = 80$

A page of congruent and similar figures

PAGE 171

Congruent triangles: B and K, C and F, E and V, I and Y, M and U, O and T.

Similar triangles (scale factor): D and W (2), J and N ($2\frac{1}{2}$), Q and A ($1\frac{1}{2}$), L and G (2), H and R (5), S and P (4), X and Z (3).

Finding sides in similar figures

PAGE 172

1 $1\frac{1}{3}$, $x = 9$ **2** 1.5, $d = 5\frac{1}{3}$

3 $\frac{2}{9}$, $r = 13.5$ **4** 1.6, $a = 3\frac{1}{8}$, $b = 3.2$

5 $\frac{4}{7}$, $u = 10.5$ **6** 1.6, $k = 10\frac{5}{8}$

7 2.5, $p = 5.6$ **8** 3, $p = 3\frac{1}{3}$, $q = 15$

9 0.4, $e = 2.5$, $f = 6$ **10** $\frac{5}{11}$, $r = 8.8$

11 $1\frac{2}{3}$, $h = 7.2$ **12** $1\frac{1}{3}$, $r = 12$

13 $1\frac{1}{3}$, $c = 7.5$, $d = 9\frac{1}{3}$ **14** 0.5, $x = 7$, $y = 4$

15 2.5, $t = 8$

Congruence and similarity crossword

PAGE 174

Across

  2 area

  4 reduction

10 included

12 original

16 image

19 triangle

24 figure

28 scale

  3 ruler

  9 similar

11 protractor

14 translation

18 enlargement

20 perimeter

25 SAS

29 reflection

Down

  1 prove

  3 ratio

  6 matching

  8 hypotenuse

15 superimpose

21 RHS

23 test

27 AAS

  2 angle

  5 congruent

  7 quadrilateral

13 rotation

17 centre

22 side

26 SSS

HOMEWORK ANSWERS

Chapter 1

Algebra 1 PAGE 8

Part A

1 $C = 2\pi r$

2 **a** 30.5 **b** 240 **c** 30

3 5 **4** 60 m^3

5 $3a$ **6** 180°

Part B

1 **a** $18y$ **b** $3d + 5e$ **c** $-35hm$

2 $5 \times m \times m \times n \times p$ **3** $Q + 6$

4 $4a^2b, 3a^2b, ba^2$ **5** $9y^2 - 18y$

6 -11

Part C

1 **a** 86 **b** -4

2 **a** $-4x^2 - 10x$ **b** $-\dfrac{3ab}{2}$ **c** $-6ab^3$

3 **a** $60x$ **b** $\dfrac{r}{3}$ **c** $10r + 6t$

Part D

1 The answer to a division **2** $2x + 2$

3 $72\ m^2$ **4** substitute

5 **a** 2, 3, 4 **b** $x - 3, x - 2, x - 1$

6 like

7 Teacher to check, for example, **c** -16

Algebra 2 PAGE 10

Part A

1 **a** $35.85 **b** $-15

c $30 **d** $1\dfrac{7}{10}$

2 2 **3** 9

4 9.9, 9.909, 9.91, 9.95 **5** 27, 243

Part B

1 **a** $-9x$ **b** $(60 - 7.4N)

2 $0, \dfrac{1}{2}, 1\dfrac{1}{2}, 2$

3 **a** $-3a + 8b$ **b** $-\dfrac{3x}{z}$

c $105r^2$

4 26

Part C

1 Teacher to check, for example, $5, x, y, 5x, y^2$

2 **a** $6a - 21$ **b** $-42v^2w + 12v^2$

3 $p(1 + 3p)$

4 12

5 **a** $5b + 17$ **b** $3xy - 3x - 4y$

6 $3mn(9n - 2m)$

Part D

1 $2n$ divides into $10n^2$ evenly

2 $a^2 + 4a$ **3** $10xy$

4 **a** expand **b** factorise

5 $4x - 6$

6 $12 \times (100 - 2) = 12 \times 100 - 12 \times 2$
$$= 1200 - 24 = 1176$$

Algebra 3 PAGE 12

Part A

1 $x = 70$ **2** 1 and 3

3 $A = \pi r^2$

4 **a** 64 **b** 144

c 3 **d** $0.70

5 $s^2 = p^2 + q^2$

Part B

1 False

2 **a** n^2 **b** $6y$

c $4 - f + 5g$

3 **a** $-16 + 10x$ **b** $4a^2 - 13a$

4 **a** $y(10 - 3y)$ **b** $\dfrac{5}{6}s(t - f)$

Part C

1 **a** $12k$ **b** $23a - 3a^2$

c $\dfrac{mn^2}{2}$

2 $65kp$

3 **a** $a^2 + 5a + 6$ **b** $2y^2 - 9y - 18$

c $-20p^2 + 54p - 10$

4 $4y - 8$

Part D

1 product **2** $\dfrac{m}{2} + 8$

3 Teacher to check, for example, if
$a = 5$, LHS $= -21$, RHS $= -21 =$ LHS

4 An expression with two terms, for example, $(y + 6)(x - 1)$

5 variable or pronumeral **6** $(90\ 500x + 62\ 000y)

7 highest common factor

Chapter 2

Pythagoras' theorem 1 PAGE 20

Part A

1 9.08 p.m. **2** 608 **3** 50 cm^2

4 220 **5** $4m(2 - m)$ **6** 76 cm

7 46 **8** $\dfrac{1}{14}$

Part B

1 4.12 **2** 73

3 **a** 9.75 **b** 11.18

4 $x = 175$ **5** 55

6 $y = 3$ **7** $m = 36$

Part C

1 **a** 17 m **b** $x = 15$

2 $\sqrt{3625}$ or $5\sqrt{145}$ **3** 14.1

4 $\sqrt{50}, \sqrt{33}, \sqrt{69}$ **5** 27.37

6 65 mm

Part D

1 **a** Length of the boundary of a shape, distance around the shape

b The longest side of a right-angled triangle

2 $c^2 = a^2 + b^2$ **3** surd

4 three squared

5

6 the sum of the square of the other two sides

Pythagoras' theorem 2 PAGE 22

Part A

1 429   **2** $\dfrac{7}{15}$ **3** $2y(3xy + 1)$

4 $-2x + 8$ **5** $24   **6** 4

7 $16.55   **8** $\angle QPR$ or $\angle RPQ$

Part B

1 a $s = 15$ **b** $f = 16$

2 $\sqrt{75}, \sqrt{41}, \sqrt{28}$

3 a $x^2 = 18^2 + 21^2$ **b** $47^2 = p^2 + 35^2$

4 a $x = 27.7$ **b** $p = 31.4$

5 $k = \sqrt{243}$ or $9\sqrt{3}$

Part C

1 a Teacher to check, $8^2 + 15^2 = 17^2$

  b Teacher to check, for example (16, 30, 34)

2 right-angled   **3** 60 m

4 82.5 mm^2

5 Teacher to check, $11^2 + 60^2 = 61^2$, right angle between the 11 and 60 sides

6 $d = 7.75$ mm   **7** $\sqrt{223}$

Part D

1 rule

2 name of ancient Greek mathematician

3 2.3 m   **4** 3.08 m

5 A root that cannot be expressed as an exact value in fraction or decimal form

6 opposite   **7** 3.19 km

8 $\sqrt{32}$ or $4\sqrt{2}$

Pythagoras' theorem revision PAGE 25

Part A

1 $\dfrac{9}{20}$ **2** 0.058   **3** 2.21

4 1   **5** $235.27   **6** 5 m^2

7 2   **8** 20

Part B

1 a $17.6^2 = a^2 + 8.3^2$ **b** $r^2 = 7.6^2 + 2.5^2$

2 a $a = 15.52$ **b** $r = 8.00$

3 yes, $9.1^2 = 3.5^2 + 8.4^2$ **4** (5, 12, 13)

5 a $y = \sqrt{1001}$ **b** $t = \sqrt{109}$

Part C

1 a Right-angled   **b** Not right-angled

2 a 23.2 cm   **b** 72 m

3 a 33.39 cm^2 **b** 216 m^2

4 4.5 units

Part D

1 hypotenuse   **2** 2.9 m

3 2.69 m

4 In a right-angled triangle, the square of the hypotenuse is equal to the sum of the squares of the other 2 sides

5 53.4 cm   **6** 46.10 km

7 A set of 3 numbers a, b, c that follows the rule $c^2 = a^2 + b^2$ (Pythagoras' theorem)

8 $\sqrt{32}$ or $4\sqrt{2}$

Chapter 3

Integers and decimals PAGE 36

Part A

1 a 4   **b** 720   **c** -15

2 $x = 116$ **3** 1, 2, 3, 6   **4** 9

5 x **6** 3.5

Part B

1 a 64%   **b** $\dfrac{16}{25}$

2 $382.50   **3** $\dfrac{2}{5}$

4 a 5832   **b** 42

  c 19

5 $\dfrac{13}{15}$

Part C

1 8.714, 8.8, 8.89

2 a -10 **b** 22.257

  c 15.059   **d** 0.022

  e 30

3 a 0.425   **b** $0.2\dot{7}$

Part D

1 2, left   **2** $22.85

3 27   **4** Samantha by 0.82 s

5 1   **6** always positive

7 $41.18

8 A decimal whose digits don't repeat but stop, such as 0.345

Fractions and percentages PAGE 38

Part A

1 2.19   **2 a** $15x + 2y$ **b** $3x^2$

3 $\dfrac{4}{5}$ **4** $A = \dfrac{1}{2}bh$ **5** $21

6 12.7   **7** $\dfrac{1}{4}$

Part B

1 $3\dfrac{6}{7}$

2 a 0.4   **b** 6.76

3 a -3 **b** 37

4 a 0.8, terminating   **b** $0.4\dot{5}$, recurring

  c 3.14159 ..., neither

Part C

1 a $\dfrac{9}{20}$ **b** 92.5%

  c 0.0875

2 a $1\dfrac{33}{40}$ **b** $1\dfrac{4}{5}$

3 $668.50   **4** $26

5 45%, 40.5%, $\dfrac{2}{5}$, $\dfrac{3}{8}$

Part D

1 275   **2** 8   **3** $140.25

4 a 6.54%   **b** 20.83%   **5** 1

6 a $81   **b** 10.7%

Percentages, ratios and rates PAGE 40

Part A

1 $4\frac{1}{2}$

2 $6xy$

3 **a** $17^2 = x^2 + 15^2$ **b** $x = 8$

4 4.15

5 $7, 6, 3, 0, -2, -5$

6 \$6

7 168

Part B

1 $0.4\dot{6}$

2 930 mL

3 Any decimal from 5.151 to 5.249

4 **a** $\dfrac{51}{125}$ **b** $\dfrac{5}{16}$

   c 64

5 \$34.16

6 Teacher to check

Part C

1 $10:3$

2 **a** \$250   **b** 26.3%

3 7 h 15 min

4 92 km/h

5 \$11 025

6 \$783

7 \$34 500

Part D

1 30 m/s

2 **a** 2 h 12 min   **b** 4.57 p.m.

3 A profit is when you sell an item for more than what you paid for it, while a loss is when you sell an item for less than what you paid for it.

4 \$5.40

5 $7:24$

6 6 min 45 s

7 9%

Chapter 4

Trigonometry 1 PAGE 52

Part A

1 a

2 $-y - 3$

3 0.0325

4 $a = 60$

5 \$102

6 yes

7 **a** 13.5   **b** 4

Part B

1 **a** a **b** $a^2 = 71^2 + 82^2$ **c** 108.47

2 $\dfrac{9}{14}$

3 **a** 29.58   **b** 8.59

4 **a** $x = 90$ **b** $d = \dfrac{7}{10}$

Part C

1 **a** 12   **b** 13   **c** 5

2 **a** cos   **b** tan   3 30° 9′

4 **a** $\dfrac{21}{29}$ **b** $\dfrac{20}{29}$

Part D

1 30

2 **a** cos   **b** tan

   c sin

3 **a** 49°   **b** 8°

4 **a** 0.63   **b** 15.07

Trigonometry 2 PAGE 54

Part A

1 $x = 72$

2 **a** \$24   **b** 4

   c $1\frac{1}{6}$ **d** 360

3 $\dfrac{1}{2}$

4 $4x^2 - 9$

5 60 m

Part B

1 **a** CA **b** BC

   c CA **d** BA

2 **a** $\dfrac{48}{73}$ **b** $\dfrac{55}{73}$

3 **a** 2.16   **b** 3.33

Part C

1 **a** 10.24   **b** 74.98

   c 14.82   **d** 39.23

2 **a** 47°   **b** 30°

3 48° 53′

4 41.9°

Part D

1 $\dfrac{\text{opposite}}{\text{hypotenuse}}$ 2 695 m   3 36° 27′

4 85.2°   5 23.67 m   6 5.46 m

7 next to   8 197.1 m

Trigonometry review PAGE 57

Part A

1 $\dfrac{5ab}{6c}$

2 **a** 24   **b** 36

3 **a** \$52.82   **b** \$36

4 **a** 12 m   **b** 6 m^2

5 $6p^2 - 9p - 6$

Part B

1 **a** 51°   **b** 27°

2 **a** 27.22   **b** 4.82

   c 55.07   **d** 204.78

3 36° 52′   4 51.4°

Part C

1 **a** 4.9   **b** 47.0

   c 6.6   **d** 24.9

2 75°   3 27.7 km

4 71.49 m   5 41°

Part D

1 **a** $\sin P = \cos Q = \dfrac{p}{r}$

   b $\tan P \times \tan Q = \dfrac{p}{q} \times \dfrac{q}{p} = \dfrac{pq}{pq} = 1$

2 **a** 26   **b** 69   **c** 26

3 36.9, 25.8, 55°   4 $\tan \theta = \dfrac{x}{x} = 1, \theta = 45°$

Chapter 5

Indices 1 PAGE 68

Part A

1 $3pq(9p - 5q)$ 2 -1

3 7.25 p.m.

4 **a** $\dfrac{1}{2}$ **b** \$14.85

5 14   6 \$51

7 $3x^2 + 8x - 16$

Part B

1 12

2 **a** 32   **b** 28.28

   c 48   **d** 8

3 a 7 to the power of 4   **b** 4

4 5

Part C

1 a y^{10} **b** $24x^7$

  c b^9 **d** $6r^4$

  e $25a^6$ **f** $\dfrac{1}{8p^3}$

  g $36m^6n^{10}$ **h** $\dfrac{-d^{10}}{e^5f^{15}}$

Part D

1 a true   **b** false

  c false

2 power or exponent   **3** multiply

4 a $\dfrac{16}{2401}$ **b** 1

5 Teacher to check, for example, $-4p^5q^2$ and $8p^2$

Indices 2   PAGE 70

Part A

1 a 210   **b** 64

  c \$8

2 0.25   **3** 82.69

4 Teacher to check   **5** 30%

6 $4d$

Part B

1 base

2 a $6x^3$ **b** $8w^{11}$

  c $2m^6$ **d** $\dfrac{3C^9}{10}$

  e $243a^{20}$ **f** $-2187p^{14}q^7r^7$

  g $\dfrac{9r^4w^3}{2}$

Part C

1 a p^{12} **b** $\dfrac{1}{y^2}$

  c 8   **d** t^{10}

  e $1\dfrac{1}{2}$ **f** $\dfrac{4}{u^3w}$

  g $\dfrac{3a}{2b}$ **h** 1

Part D

1 a false   **b** true

  c true

2 a $-\dfrac{1}{125}$ **b** $\dfrac{1}{16}$

3 a subtract   **b** p^nq^n

4 Teacher to check, for example, $18y^6z^2$ and $3y^2z$

Indices 3   PAGE 72

Part A

1 $0.\dot{3}$ **2** -216 **3** $f = 40$

4 $\dfrac{9}{20}$ **5** $4d + 2p$ **6** 0

7 no   **8** $-15a + 6$

Part B

1 a $\dfrac{k^2}{3w}$ **b** $8p^{24}$ **c** $-\dfrac{1}{243}$

  d $\dfrac{3}{8}$ **e** $-36m^9n^2$

2 a false   **b** false   **3** $n = -2$

Part C

1 a 5.7×10^{-8} **b** 4.38×10^8

2 26.1   **3** 0.008

4 a 50 300 000   **b** 0.000 075

5 a 3.6×10^{12} **b** 230

Part D

1 5 800 000   **2 a** 3×10^8 **b** 8×10^{-5}

3 a 10, numbers   **b** end   **c** itself

4 a 12 is not a number between 1 and 10

  b 12 000, 1.2×10^4

Indices revision   PAGE 74

Part A

1 $15p^2$ **2** 12   **3 a** 270

  b -27 **c** 106   **4** y

5 $4x^2 + 20x + 25$ **6** 5 h 45 min

Part B

1 a 0.079

2 a 5.149×10^{-4} **b** 5.23×10^3

  c 8.1×10^0

3 21 000   **4** 1.225×10^{-5}

5 3   **6** $n = -2$

Part C

1 a 5   **b** $\dfrac{2}{y^3}$

  c $6h^3$ **d** $2u^9$

  e $\dfrac{9a^3}{b^2c^2}$

2 37 870 000   **3** 4.76×10^5

4 8.0×10^{10}

Part D

1 1   **2** 1×10^{12}

3 a true   **b** false

4 a multiply   **b** beginning

  c 0

5 152 000 000 km

Chapter 6

Geometry 1   PAGE 84

Part A

1 a 4   **b** 4.5   **2** 5 : 2

3 \$20   **4** 64 cm^3 **5** \$168

6 $5k^5$ **7** $14x + 4$

Part B

1 a acute   **b** right   **c** obtuse

  d reflex

2 a 180°   **b** 180°   **c** 180°

  d 60°

Part C

1 a $b = 60$ **b** $d = 75$ **c** $r = 75$

  d $a = 40$ **e** $n = 132$ **f** $c = 140$

  g $k = 70$ **h** $m = 68$

Part D

1 a the two interior opposite angles

  b equal   **c** 90°

  d 360°

2

 or

3 0   **4** 2

5 alternate angles are not equal

Geometry 2

Part A

1 $b^2 = a^2 + c^2$ **2 a** 340   **b** -10

3 10   **4** 120 m^2 **5** $\dfrac{7}{25}$

6 $\dfrac{9x^{10}}{16}$ **7** sin

Part B

1 a $a = 65$ **b** $u = 60$ **c** $b = 100$

   d $p = 32$ **e** $c = 30$ **f** $y = 45$

   g $r = 50$ **h** $x = 152$

Part C

1 a rectangle   **b** rhombus

   c square   **d** kite

2 a $w = 50$ **b** $k = 20$

   c $y = 30$ **d** $n = 35$

Part D

1 a A quadrilateral with two pairs of parallel sides

   b opposite angles

   c diagonals bisect each other

   d 360°

2 $n = l + m$

3 a square

   b $l = 45$, diagonals bisect the angles of the square

4 a   **b** 0

Chapter 7

Equations 1

Part A

1 29.3   **2** $924   **3** 0.0905

4 kite, 1   **5** $-x + 5y$ **6** 32 cm^2

7 t^{11} **8** 0.002

Part B

1 a $n = 40$ **b** $y = 5$ **c** $r = 27$

2 a $4x + 12$ **b** $8d - 56$ **c** $-5a + 10$

3 a -34 **b** 12

Part C

1 a $u = 18$ **b** $z = -9$ **c** $h = 5$

   d $b = 45$ **e** $a = 15$ **f** $p = -5$

   g $y = -8$ **h** $x = 7$

Part D

1 a $5x + 2$ **b** $x = 11$

2 Teacher to check, for example,

   a Subtract $3x$ from both sides   **b** Add 2 to both sides

   c Divide both sides by 3

3 Teacher to check, for example, $4x - 10 = 10$.

---

4 $5y = 0$

5 Substitute the solution back into the equation to show that both sides are equal

Equations 2

Part A

1 a 0.874   **b** 87.4%

2 $x = 10$ **3** 216

4 89.9, 89.901, 89.918, 89.92   **5** 10 000

6 4.95×10^5 **7** $2l + 2w$ or $2(l + w)$

Part B

1 a $a = 7$ **b** $x = -\dfrac{2}{3}$

   c $h = 2\dfrac{3}{8}$ **d** $r = -13$

   e $a = 6$ **f** $p = 8$

   g $t = 2$ **h** $u = -6\dfrac{1}{2}$

Part C

1 a $w = 4$ **b** $d = 5\dfrac{1}{3}$

   c $m = 2\dfrac{1}{11}$ **d** $x = 5$

   e $a = 7$ **f** $y = 36$

2 length = 14 m, width = 7 m   **3** 34 and 36

Part D

1 84°, 58°, 38°   **2** $v = 12.5$ **3** solution

4 8   **5** 10

Equations revision

Part A

1 $5y(3 - x)$ **2** $\dfrac{5}{24}$ **3** 125 cm^3

4 8.138   **5** $x = 120$ **6** $175

7 $3x^2 - 17x + 20$ **8** Rose $800, Khalid $1000

Part B

1 a $y = 5$ **b** $x = 8$ **c** $n = 3$

   d $p = 21$ **e** $w = 1$ **f** $b = -\dfrac{1}{2}$

   g $k = -11$ **h** $s = 1\dfrac{1}{3}$

Part C

1 a $x = -19$ **b** $f = 1\dfrac{1}{4}$

2 a 23 m   **b** 414 m^2

3 26, 27, 28

4 a $F = 23$ **b** $C = 30$

Part D

1 a variable or pronumeral

   b Teacher to check, for example, expand the LHS

   c Teacher to check, for example, add 32 to both sides

2 $p = -5\dfrac{1}{5}$ **3** $w = 7$

4 9   **5** 309

Chapter 8

Earning money

Part A

1 45   **2** 11, 13, 17 or 19   **3** 30 cm^2

4 $4.60   **5** $-6a - 18$ **6** $22

7 CAD or DAC **8** 6

Part B

1 26   **2** 0.175
3 a $1359.35   **b** $28 450.53
4 8 h 15 min   **5** 1.5%
6 $71.40   **7** $644

Part C

1 $1260   **2** $1213.12   **3** $965.52
4 3.8%   **5** $1104.60   **6** $79 739.40
7 a $900   **b** $630

Part D

1 pay for work based on a fixed yearly amount
2 Annual pay = $3850 × 12 months = $46 200,
   Weekly pay = $46 200 ÷ 52.18 ≈ $885.40
3 a 21   **b** 7
   c $669.10
4 a $26.04   **b** $1299.92
5 annual leave loading

Chapter 9

Data 1

PAGE 118

Part A

1 cylinder   **2** $x = 22\frac{1}{2}$ **3** $25.50
4 $x = 60$ **5** -7 **6** $\dfrac{25b^2}{a}$
7 $1 - p$ **8** 50 cm^3

Part B

1 a 4.9   **b** 5   **c** 4.5
   d 14   **e** 15
2 a 22   **b** 6   **c** 4

Part C

1 a

Score, x	Frequency, f	fx
3	5	15
4	3	12
5	7	35
6	6	36
7	4	28
Totals:	**25**	**126**

   b 25   **c** 5.04   **d** 5
   e

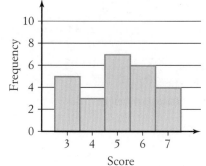

2 a Piglets, higher scores shown on stem-and-leaf plot
   b 42   **c** 41

Part D

1 a most common data value(s)
   b extreme value, different to the other values
   c column graph showing frequencies of numerical data
2 mean
3 a location, centre or central tendency   **b** mean
   c cluster
4 Teacher to check, for example, 2, 3, 17, 18, 20.

Data 2

PAGE 120

Part A

1 54 cm^2 **2** $x = -5\frac{1}{5}$
3 $p = 128$, angles on a straight line
4 Teacher to check: 4 numbers with a sum of 28, range
   of 9, for example, 2, 6, 9, 11.
5 isosceles triangle, 1   **6** $\dfrac{13a}{12}$
7 $\dfrac{9\pi}{4}$ cm^2

Part B

1 a 6 and 7   **b** 7
   c 4   **d** 6
2 a 8, 26, 46, 51, 52   **b** 1.5
   c 2

Part C

1 a 19   **b** positively skewed
   c 51   **d** 50's
2 a Melbourne   **b** Brisbane
   c Brisbane   **d** Melbourne

Part D

1 a

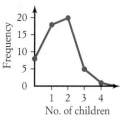

   b positively skewed

2 a bunched/grouped   **b** symmetrical
   c positively
3 a outlier   **b** mode
   c polygon

Data 3

PAGE 122

Part A

1 $4\frac{5}{6}$ **2** $p = 20$ **3** $\frac{2}{3}$
4 $2xy^2 + 14x^2y$ **5** $304   **6** 20 m^2
7 10 m   **8** $3x^2 - 22x - 45$

Part B

1 a 1 and 7   **b** symmetrical
   c 6   **d** 4
2 a i girls **ii** boys **iii** girls   **b** 13.3%

Part C

1 a sample   **b** census
   c census
2 a categorical   **b** discrete numerical
   c continuous numerical   **d** categorical

3 Biased towards visitors to that website, only interested or strong-viewed people vote, people can vote more than once, sample is not random

Part D

1 a mode   **b** mean

   c mean

2 Tail points to the left or lower values, most values at right or higher values

3 a frequency   **b** census

   c continuous   **d** bias

Chapter 10

Area 1 PAGE 130

Part A

1 7.35 p.m.   **2** $\dfrac{2}{3}$ **3** $x^2 = d^2 + r^2$

4 $5xy(2x - 3y)$ **5** $7:4$ **6** $\dfrac{1}{81}$

7 $7w$ **8** $x = 5\dfrac{2}{5}$

Part B

1 a trapezium   **b** triangle

   c rhombus   **d** circle

   e parallelogram

2 a 183 cm   **b** 31.2 m

3 31 m^2

Part C

1 a 160 cm   **b** 960 cm^2

2 a 81 cm^2 **b** 24.8 cm^2

   c 18 m^2 **d** 13.5 cm^2

   e 150 mm^2 **f** 40.5 cm^2

Part D

1 a square centimetre

   b The area of a square of length 1 cm

   c 100

2 perimeter   **3** 4 m

4 a $16x + 4y$ **b** $16x^2 + 8xy$

5 36.2 cm

Area 2 PAGE 133

Part A

1 a   **b** 5

2 3737   **3** 68%, $\dfrac{2}{3}$, 0.65, 0.6

4 $x^2 - 4x - 60$ **5** $p = 70$

6 5 h 30 min   **7** 8 m^3

Part B

1 a $\sqrt{113}$ **b** 25.63 m

2 a 80 m^2 **b** 225 cm^2

   c 26.65 m^2 **d** 10.5 cm^2

3 46 m   **4** $A = \pi r^2$

Part C

1 a 50.27 cm   **b** 37.70 mm

2 a 201.06 cm^2 **b** 113.10 mm^2

3 a 15.4 mm   **b** 16.5 m

4 a 14.5 mm^2 **b** 16.4 m^2

Part D

1 It does not have an exact fraction or decimal value

2 a 39.27 m^2 **b** 164.55 m^2

3 a circumference   **b** diameter

   c sector

4 a 125.7 cm^2 **b** 42.1 m^2

Surface area PAGE 136

Part A

1 a 572.8   **b** 26

2 $x = 65$, corresponding angles on parallel lines

3 mode   **4** 24

5 $x = 1$ **6** $-6ab - 6ab^2$

7 $800

Part B

1 a 39.27 m^2 **b** 201.06 cm^2

   c 76.27 mm^2 **d** 205.25 m^2

2 a 25.71 m   **b** 57.13 cm

   c 34.85 mm   **d** 57.32 m

Part C

1 a 210 cm^2 **b** 2318.50 cm^2

   c 540 m^2 **d** 471.32 cm^2

   e 1176 mm^2 **f** 12 666.90 m^2

2 2400 m^2

Part D

1 a 3   **b** 2 circles, 1 rectangle

2 $2\pi rh$ **3** 180 m^2

4 The sum of the area of the faces of the solid

5 530 cm^2 **6** 3 cm

Volume PAGE 139

Part A

1 $40p^4 r$ **2** $6\dfrac{2}{3}$ **3** -65

4 y-axis   **5** $y = 122$ **6** $\dfrac{a}{42}$

7 a 21.5   **b** 36

Part B

1 a 55.42 m^2 **b** 25.72 mm^2

   c 42 m^2 **d** 56.55 cm^2

2 a 2170 cm^2 **b** 100.78 mm^2

   c 8301.66 mm^2 **d** 387.54 cm^2

Part C

1 a $V = \pi r^2 h$ **b** $SA = 2\pi r^2 + 2\pi rh$

2 a 472.90 m^3 **b** 168 cm^3

   c 7379.60 mm^3 **d** 712.3995 m^3

3 a volume of a prism

   b A = area of base, h = height

Part D

1 circle

2 a 1.37 L   **b** 706 cm^2

3 a cubic metre

   b the volume of a cube of length 1 m

   c 1 000 000

4 a 13 254 cm^3 **b** 81 628 cm^3

Chapter 11

Coordinate geometry 1
PAGE 148

Part A

1 $10 : 9$ **2** $22.30   **3** 24 cm^2

4 53   **5** 7.05   **6** 9.51×10^{-3}

7 a 8.375   **b** 13

Part B

1 $\dfrac{3}{5}$

2

x	-2	-1	0	1	2
y	-5	-3	-1	1	3

3 a 3   **b** $2\dfrac{1}{2}$

4 a 4   **b** 7

   c -3

5 $p = 61$

Part C

1 a $\sqrt{117}$ or $3\sqrt{13}$ **b** $(1, \dfrac{1}{2})$

   c $-\dfrac{3}{2}$

2

(a line increasing at a 45° angle)

3 a $(4, 6)$ or $(0, 8)$ **b** 4.5 units

   c $(2, 7)$ **d** $\dfrac{1}{2}$

Part D

1 a The point at the centre of an interval, halfway between the endpoints

   b Teacher to check: finding averages of the x- and y-coordinates or using the formula

   c Teacher to check: because the line slopes downwards

   d Rise is how far the interval goes up, run is how far the interval goes across/right, for PQ, rise $= 10$, run $= 12$.

2 Pythagoras' theorem

3 gradient $= \dfrac{0}{4} = 0$, because it is a horizontal/flat line

Coordinate geometry 2
PAGE 150

Part A

1 a 72   **b** $\dfrac{5}{12}$

   c $\dfrac{16}{25}$ **d** 420

2 6.8   **3** $153

4 $x = 110$ **5** 2 and 3

Part B

1 a 4.5 units   **b** $(2, 3)$

   c 2

2 a $-\dfrac{2}{3}$ **b** $\dfrac{1}{3}$

3 a $\sqrt{356}$ or $2\sqrt{89}$ **b** $(2, -1)$

   c $-\dfrac{5}{8}$

Part C

1 a

x	-2	-1	0	1	2
y	1	$1\dfrac{1}{2}$	2	$2\dfrac{1}{2}$	3

b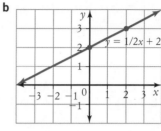

   c 2   **d** $\dfrac{1}{2}$

2 $p = 75$

3 a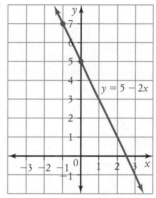

   b no   **c** $x = 1\dfrac{1}{2}$

Part D

1 a 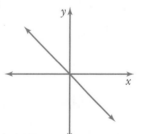   **b** $y = -x$

2 x-intercept

3 horizontal line with y-intercept 2

4 k

5 a $D = 0.8t$ or $\dfrac{4t}{5}$ **b** 12.5 min

6 Teacher to check, for example, $\dfrac{\text{rise}}{\text{run}}$

Coordinate geometry 3
PAGE 152

Part A

1 1.21 cm^2 **2** 0.91

3 $920   **4** $xy + 6x$

5 $y = 80$ **6** 4

7 $A = \dfrac{1}{2}xy$ **8** 12

HW HOMEWORK ANSWERS

Part B

1 a

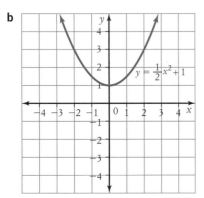

b -3　　　　　**c** $1\frac{1}{2}$　　　　　**d** 2

2 a $M = 2.5V$　　**b** 1510 g　　　**c** 400 cm^3

3 $y = -\frac{1}{2}x$

Part C

1 a

x	-2	-1	0	1	2
y	3	$1\frac{1}{2}$	1	$1\frac{1}{2}$	3

b

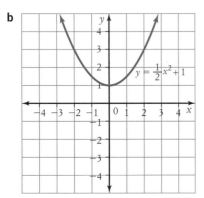

c 1　　　　**d** parabola　　　**e** concave up

Part D

1 a $(0, 1)$　　　**b** vertex　　　**2** $x^2 + y^2 = 25$

3 a $y = -2$　　**b** $y = \frac{2}{5}x + 2$　　**4** -3

5 radius of a circle　　**6** y-intercept of a line

Coordinate geometry revision　　　　PAGE 154

Part A

1 a 70　　　　　　**b** 10 000 000 or 10^7　　**2** 78.5 cm^2

3 0　　　　　　　**4** $4x + 12$　　　　**5** \$1400

6 9　　　　　　　**7** 32 m

Part B

1 a

x	-2	-1	0	1	2
y	-1	2	3	2	-1

b

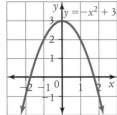

c -1.7 and 1.7

d $(0, 3)$　　　　　　**e** concave down

2

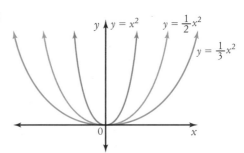

3 $x^2 + y^2 = 81$

Part C

1 yes

2 a $(-1, 2)$　　　　　　　**b** 10 units

c $-\frac{4}{3}$

3 a gradient $= -1$, y-intercept $= -2$

b

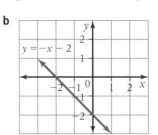

4 165.6

Part D

1 -2, or any value less than -1

2 a parabola, down　　　　**b** m

c midpoint　　　　　　　**d** rise, run

3 Teacher to check, for example,

a $y = 3x^2 - 1$　　　　**b** $y = 3x - 1$

4 $(0, 0)$

Chapter 12

Probability 1　　　　　　　　PAGE 162

Part A

1 0.68　　　　　　　　**2** 12.4

3 $\frac{p^2}{4y^2}$　　　　　　　**4** $\frac{b}{a}$

5 horizontal　　　　　**6** -6

7 \$42.50　　　　　　**8** 28.8 m^2

Part B

1 a {head, tail}　　　　**b** {1, 2, 3, 4, 5, 6}

c {hearts, clubs, diamonds, spades}

2 0.8　　　　　　　　**3** 38.3%

4 a even chance　　　　**b** Teacher to check

c Teacher to check

Part C

1 a $\frac{1}{6}$

b rolling a number other than 4

c 1

2 a i 5　　　　　　　　**ii** 5

b $\frac{7}{10}$

3 a 0.58　　　　**b** 139

Part D

1 $\frac{4}{5}$

2 **a** the set of all possible outcomes of a chance experiment

  b where every possible outcome is equally likely

3 the probability of a tossed coin not showing heads (showing tails)

4 33%   **5** outcome

6 $\frac{3}{13}$

Probability 2 PAGE 164

Part A

1 $2.14\dot{6}$ **2** 21

3 sloping downwards from left to right

4 $\frac{a}{c}$ **5** $8y(2xy - k)$

6 5.55 a.m.

7 No   **8** $x = 60$

Part B

1 $\frac{2}{7}$

2

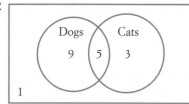

    Dogs   Cats
    9 5 3
    1

3 **a** rolling the number 4 or less **b** $\frac{2}{3}$

4 **a** $\frac{3}{10}$ **b** $\frac{2}{9}$

5 7

Part C

1 **a** 13   **b** 94

2 **a** {HH, HT, TH, TT} **b i** $\frac{1}{4}$ **ii** $\frac{1}{2}$ **iii** $\frac{3}{4}$

3 **a** $\frac{9}{17}$ **b** $\frac{11}{17}$

Part D

1 **a**

		Die					
		1	2	3	4	5	6
Coin	H	H1	H2	H3	H4	H5	H6
	T	T1	T2	T3	T4	T5	T6

b i $\frac{1}{12}$ **ii** $\frac{1}{6}$

2 All items or categories   **3** $\frac{9}{16}$

4 Teacher to check, for example, rolling two dice

5 {13, 17, 18, 31, 37, 38, 71, 73, 78, 81, 83, 87}

Probability revision PAGE 166

Part A

1 The lengths of the diagonals of the rhombus

2 **a** $9a^{10}$ **b** 3

3 $x = 45$

---

4 **a** $13   **b** $\frac{5}{13}$

5 3.03   **6** 8 and 10

Part B

1 **a i** $\frac{1}{5}$ **ii** $\frac{2}{5}$ **iii** $\frac{3}{5}$

  b 20

2 **a**

	Saturday	Sunday	Outcomes
		R	R R
	R		
		$\overline{R}$	R $\overline{R}$
		R	$\overline{R}$ R
	$\overline{R}$		
		$\overline{R}$	$\overline{R}$ $\overline{R}$

  b i $\frac{1}{4}$ **ii** $\frac{1}{2}$ **iii** $\frac{3}{4}$

Part C

1 **a** $\frac{2}{13}$ **b** $\frac{5}{13}$

2 **a** 0.222   **b** 0.744

3 **a** 18   **b** 28

  c 55   **d** 13

Part D

1 **a i** 25.625% **ii** 3.125% **iii** 96.875%

2 **a**

		1st die					
	–	1	2	3	4	5	6
2nd die	1	0	1	2	3	4	5
	2	1	0	1	2	3	4
	3	2	1	0	1	2	3
	4	3	2	1	0	1	2
	5	4	3	2	1	0	1
	6	5	4	3	2	1	0

  b i $\frac{1}{18}$ **ii** $\frac{1}{2}$ **iii** $\frac{4}{9}$

Chapter 13

Congruent figures PAGE 176

Part A

1 18 h   **2** $12   **3** false

4 $x = 110$ **5 a** 26   **b** 0.3452

6 $\frac{5a^4}{b}$ **7** 0.002

Part B

1 A and E, C and F

2 **a** rotation   **b** JK

  c $\angle M$ **d** $KLMJ$

3 **a** true   **b** true

Part C

1 **a** $\angle T = \angle W$, $\angle S = \angle W$, $\angle R = \angle U$

  b $TR = WU$, $TS = WV$, $RS = UV$

2 **a** SAS   **b** AAS

3 a SSS   **b** 90°
  c cross at right angles   **d** no

Part D
1 a side   **b** hypotenuse
2 a SAS   **b** ACB
3 a $\angle A = \angle C,\ \angle ADB = \angle CDB,\ \angle ABD = \angle CBD$
  b angles, angles

Similar figures 1 PAGE 178

Part A
1 8   **2** 6 h
3 $6x + 5x^2$ **4** $x = 35$
5 Volume of a cylinder   **6** 0.5
7 $4p^2 + 31p - 8$ **8** 150 cm²

Part B
1 AAS   **2** included angle
3 In $\triangle MNP$ and $\triangle ZXY$:
    $MN = ZX = 22$ m
    $NP = XY = 18$ m
    $MP = ZY = 11$ m
    $\therefore \triangle MNP \equiv \triangle ZXY$ (SSS)

4 a, b   **c** bisect the angles of
                     the rhombus

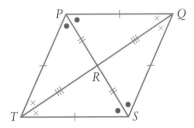

Part C
1 a 2   **b** $\dfrac{1}{4}$
2 a 128 mm   **b** 6 mm
3 a 2   **b** $y = 10$
4 8.4 m   **5** $p = 2.5$

Part D
1 a false   **b** true
  c true   **d** false
2 smaller, a reduction
3 a SAS   **b** 1.5
  c $\angle B$ **d** RBX

Similar figures 2 PAGE 181

Part A
1 a $\dfrac{11}{21}$ **b** 1584

2 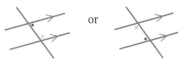 or

3 $2a - 8b$ **4** mode
5 $x, x + 2, x + 4$
6 traffic light showing green or amber
7 $560   **8** $-7y(x + 3y)$

Part B
1 a equal   **b** in the same ratio

2 a 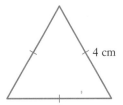   **b**
3 a $x = 33\dfrac{1}{3}$ **b** $p = 6$
4 a $\angle W$ **b** TU

Part C
1 a corresponding angles on parallel lines
  b equiangular / AA
  c $x = 9\dfrac{1}{3}$
2 a vertically opposite angle
  b PON
  c 15 cm   **d** $\dfrac{4}{25}$
3 $p = 10$

Part D
1 1   **2** is similar to
3 3.6 m   **4** SSS, SAS, AAS, RHS
5 each length is smaller by $\dfrac{1}{3}$
6 a SAS   **b** $\dfrac{2}{5}$
  c 6.4 cm

Congruent and similar figures PAGE 183

Part A
1 a $240   **b** 72
2 $5m + 2$ **3** $5\dfrac{3}{4}$
4 2   **5** Area of a trapezium
6 $m = 105$ **7** x^5

Part B
1 a In $\triangle PSR$ and $\triangle PQR$:
    $PS = PQ$ (given)
    $\angle SPR = \angle QPR = 50°$
    PR is common
    $\therefore \triangle PSR \equiv \triangle PQR$ (SAS)
  b $x = 9, y = 35$
  c kite, 2 pairs of equal adjacent sides
2 a equiangular/AA   **b** $x = 4$

Part C
1 a RHS   **b** AAS
2 a $\angle ABE = \angle ACD$, or $\angle AEB = \angle ADC$
  b equiangular/AA   **c** $y = 12$
3 a RHS   **b** SUT
  c $x = 20$

Part D
1 a congruent, SAS   **b** congruent, RHS
  c not congruent
2 a Matching sides are in the same ratio:
    $\dfrac{6}{8} = \dfrac{5.25}{7} = \dfrac{7.5}{10} = \dfrac{3}{4}$
  b FGE
3 a true   **b** false